Health Care Engineering: Part I

Clinical Engineering and Technology Management

Synthesis Lectures on Biomedical Engineering

Editor
John D. Enderle, *University of Connecticut*

Lectures in Biomedical Engineering will be comprised of 75- to 150-page publications on advanced and state-of-the-art topics that spans the field of biomedical engineering, from the atom and molecule to large diagnostic equipment. Each lecture covers, for that topic, the fundamental principles in a unified manner, develops underlying concepts needed for sequential material, and progresses to more advanced topics. Computer software and multimedia, when appropriate and available, is included for simulation, computation, visualization and design. The authors selected to write the lectures are leading experts on the subject who have extensive background in theory, application and design.

The series is designed to meet the demands of the 21st century technology and the rapid advancements in the all-encompassing field of biomedical engineering that includes biochemical, biomaterials, biomechanics, bioinstrumentation, physiological modeling, biosignal processing, bioinformatics, biocomplexity, medical and molecular imaging, rehabilitation engineering, biomimetic nano-electrokinetics, biosensors, biotechnology, clinical engineering, biomedical devices, drug discovery and delivery systems, tissue engineering, proteomics, functional genomics, molecular and cellular engineering.

Health Care Engineering: Part II: Research and Development in the Health Care Environment
Monique Frize
November 2013

Health Care Engineering: Part I: Clinical Engineering and Technology Management
Monique Frize
November 2013

Computer-aided Detection of Architectural Distortion in Prior Mammograms of Interval Cancer
Shantanu Banik, Rangaraj M. Rangayyan, J.E. Leo Desautels
January 2013

Content-based Retrieval of Medical Images: Landmarking, Indexing, and Relevance Feedback
Paulo Mazzoncini de Azevedo-Marques, Rangaraj Mandayam Rangayyan
January 2013

Chronobioengineering: Introduction to Biological Rhythms with Applications, Volume 1
Donald McEachron
October 2012

Medical Equipment Maintenance: Management and Oversight
Binseng Wang
October 2012

Fractal Analysis of Breast Masses in Mammograms
Thanh M. Cabral, Rangaraj M. Rangayyan
October 2012

Capstone Design Courses, Part II: Preparing Biomedical Engineers for the Real World
Jay R. Goldberg
September 2012

Ethics for Bioengineers
Monique Frize
2011

Computational Genomic Signatures
Ozkan Ufuk Nalbantoglu and Khalid Sayood
2011

Digital Image Processing for Ophthalmology: Detection of the Optic Nerve Head
Xiaolu Zhu, Rangaraj M. Rangayyan, and Anna L. Ells
2011

Modeling and Analysis of Shape with Applications in Computer-Aided Diagnosis of Breast Cancer
Denise Guliato and Rangaraj M. Rangayyan
2011

Analysis of Oriented Texture with Applications to the Detection of Architectural Distortion in Mammograms
Fábio J. Ayres, Rangaraj M. Rangayyan, and J. E. Leo Desautels
2010

Multimodal Imaging in Neurology: Special Focus on MRI Applications and MEG
Hans-Peter Müller and Jan Kassubek
2007

Estimation of Cortical Connectivity in Humans: Advanced Signal Processing Techniques
Laura Astolfi and Fabio Babiloni
2007

Brain–Machine Interface Engineering
Justin C. Sanchez and José C. Principe
2007

Introduction to Statistics for Biomedical Engineers
Kristina M. Ropella
2007

Capstone Design Courses: Producing Industry-Ready Biomedical Engineers
Jay R. Goldberg
2007

BioNanotechnology
Elisabeth S. Papazoglou and Aravind Parthasarathy
2007

Bioinstrumentation
John D. Enderle
2006

Fundamentals of Respiratory Sounds and Analysis
Zahra Moussavi
2006

Advanced Probability Theory for Biomedical Engineers
John D. Enderle, David C. Farden, and Daniel J. Krause
2006

Intermediate Probability Theory for Biomedical Engineers
John D. Enderle, David C. Farden, and Daniel J. Krause
2006

Basic Probability Theory for Biomedical Engineers
John D. Enderle, David C. Farden, and Daniel J. Krause
2006

Health Care Engineering: Part I: Clinical Engineering and Technology Management
Monique Frize, P. Eng., O.C.

ISBN: 978-3-031-00529-9   paperback
ISBN: 978-3-031-01657-8   ebook

DOI 10.1007/978-3-031-01657-8

A Publication in the Springer series
SYNTHESIS LECTURES ON BIOMEDICAL ENGINEERING #50
Series Editor: John D. Enderle, University of Connecticut

Series ISSN 1930-0328 Print   1930-0336 Electronic

Health Care Engineering: Part I

Clinical Engineering and Technology Management

Monique Frize
Carleton University and University of Ottawa

SYNTHESIS LECTURES ON BIOMEDICAL ENGINEERING #50

ABSTRACT

The first chapter describes the health care delivery systems in Canada and in the U.S. This is followed by examples of various approaches used to measure physiological variables in humans, either for the purpose of diagnosis or monitoring potential disease conditions; a brief description of sensor technologies is included. The function and role of the clinical engineer in managing medical technologies in industrialized and in developing countries are presented. This is followed by a chapter on patient safety (mainly electrical safety and electromagnetic interference); it includes a section on how to minimize liability and how develop a quality assurance program for technology management. The next chapter discusses applications of telemedicine, including technical, social, and ethical issues. The last chapter presents a discussion on the impact of technology on health care and the technology assessment process.

KEYWORDS

Health care Canada and U.S., clinical engineering, sensor technologies, patient safety, telemedicine, technology assessment.

Contents

Preface

The material for this Lecture Series arises from 20 years of teaching a course entitled "Health Care Engineering." My background was an undergraduate degree in Electrical Engineering at the University of Ottawa, a Master of Philosophy in Electrical Engineering in Medicine at Imperial College of Science and Technology in London (U.K.) and a doctorate at Erasmus Universiteit in Rotterdam (The Netherlands) in clinical engineering. After the Master degree in London in 1970, I taught two courses—bioinstrumentation and biomedical statistics—at Université du Québec à Montréal for one term, then worked as a clinical engineer at Notre-Dame Hospital for eight years. In this role, I developed the medical equipment management program and realized quickly how engineers have a significant impact on the quality, effectiveness and safety of health care delivery. In July 1979, I was appointed Head of the Regional Clinical Engineering Services for seven hospitals in North-Eastern New-Brunswick. In 1989, after completing the doctoral degree, I was appointed Professor in Electrical/Biomedical Engineering at the University of New Brunswick, where I developed the health care engineering course (Part I). In 1996, as Professor at Carleton University and the University of Ottawa, I developed a second course for graduate students (Part II).

This book provides a basis on which other specialized courses can be built, such as signal processing, medical imaging, data mining, processing, etc. It is useful for hospital administrators, clinical engineers, and biomedical technicians who hold a direct or indirect role for technology management in their institution. References and additional readings are provided for readers who wish to have a more in-depth knowledge about each of the topics discussed.

Part II of this two-part series presents a brief discussion on the occurrence of adverse events and medical errors and how information technology can help to reduce these. Chapter 2 discusses issues related to electronic medical records. The next three chapters present the four steps in the knowledge management process and illustrate these through examples taken from the perinatal clinical environment. These steps include data acquisition, storage, and retrieval; knowledge discovery; knowledge translation; and knowledge integration and sharing. Chapter 6 is a short discussion on clinical trials and usability studies.

Both Part I and Part II of this Health Care Engineering book consolidate material that supports a course on medical technology management and research and development of technologies in a health care environment. It can be useful for anyone involved in design, development, or research, whether in industry, hospitals, government, or in universities and colleges. It is not intended to cover all topics in depth but rather to provide an overview of the subjects and sources where additional information can be found.

It is helpful here to explain a few of the words used in the book. The term "biomedical technician" is used instead of biomedical technologist to differentiate between the medical functions carried out by medical imaging technologists or respiratory technologists. Some biomedical technical employees have four years of technical education, which would label them as technologists, but the term "technician" is used here to refer to people involved in the repair and maintenance of equipment, as opposed to people operating it.

Another term that conjures up debates in the field is "medical device." WHO defines medical device as an: "article, instrument, apparatus or machine that is either used in the prevention, diagnosis or treatment of illness or disease, or used for detecting, measuring, restoring, correcting or modifying the structure or function of the body for some health purpose" (WHO, http://www.who.int/medical_devices/technology_diffusion_24Jan12.pdf; last accessed September 2013). In this book, the term "equipment" is used most frequently, but the term medical device is also used in some places.

I am grateful to all the students who have provided interesting discussions in their essays and class discussions. They have contributed positively to an improvement of the course over the years. My hope is that they become responsible engineers, who design and develop technological solutions with people, society, and our world in mind.

CHAPTER 1

The Health Care System in North America (Canada and U.S.)

This chapter summarizes some of the important elements of the health care system in Canada and the U.S. and is not intended to be comprehensive. Rather, the intent is to familiarize readers with the main structure and funding of health care in these two countries. The latter part of the chapter provides a brief description of health care institutions, to help engineers and technicians interface with these organizations and their human resources while performing their biomedical work.

Canada and the U.S. had a similar health care system in the early 1960s, but in the past 50 years funding models in both countries have differed substantially. Canada's universal health care services are covered to the level of about 70% by government. The Canada Health Act states that all citizens are to be fully insured and will not have to disburse user fees for all medically necessary hospital and physician care. About 91% of hospital expenditures and 99% of physician services are financed by the public sector. However, ophthalmology and dental services account for much of the private expenditures in Canada. The U.S. has a mixed public-private system and is one of two Organization for Economic Co-operation and Development (OECD) countries not to have some form of universal health coverage, the other being Turkey.

1.1 CANADA

Health Canada defines its role with regards to the Canadian Health care system:

> Roles and responsibilities for Canada's health care system are shared between the federal and provincial-territorial governments. Under the *Canada Health Act* (CHA), the federal health insurance legislation, criteria and conditions are specified that must be satisfied by the provincial and territorial health care insurance plans in order for them to qualify for their full share of the federal cash contribution, available under the Canada Health Transfer (CHT). Provincial and territorial governments are responsible for the management, organization and delivery of health services for their residents (**Health Canada, 2010**).

There are 13 health systems in Canada, with common features and variations within the 10 provinces and 3 territories. The Government of Canada established basic national standards and rules and contributes a share of the health care delivery costs carried out by the provinces, while the latter oversee, plan, manage, and pay the lion share of the costs. What Canadians used to call Medicare consisted of the two historical agreements between the Government of Canada and the provinces to cover all hospital costs and doctors' visits. But with time, Medicare eventually incorporated all health expenditures undertaken by each of the provinces. Exceptions that remained were drugs, homecare, and dental care. In addition to the governments, there are other players: professional associations, industries (pharmaceutical and medical devices), service agencies, Non-Governmental Organizations, the Conference Board of Canada, the World Health Organization (WHO) in Geneva, as well as their regional components such as the Pan American Health Organization (PAHO) in the Americas (Washington) and a Regional Office for Europe (Copenhagen). The Organization for Economic Cooperation and Development (OECD) with Headquarters in Paris also studies the health systems of its members. The World Bank, located in Washington, Paris, and Tokyo, the International Monetary Fund in Washington, and the International Labour Office (ILO) in Geneva are other organizations involved in health care issues.

WHO states "Health systems consist of all the people and actions whose primary purpose is to improve health. They may be integrated and centrally directed, but often they are not…" (WHO, 2000). The World Health Organization provides some comments on health systems in general:

> They have contributed enormously to better health, but their contribution could be greater still, especially for the poor. Failure to achieve that potential is due more to systemic failings than to technical limitations. It is therefore urgent to assess current performance and to judge how health systems can reach their potential (**World Health Organization, 2000**).

In Canada, the health system consists of the interactions between three major players: Health Canada and the Federal Government; the provincial Ministries of Health and their respective governments; and organized medicine. No one is really in charge of the system, which rests on the constantly renegotiated equilibrium between these key players. The system in Canada is not "socialized medicine" but rather a "social insurance" system, as doctors are in the private sector and Canadian hospitals are controlled by private boards and/or regional health authorities, rather than being part of the government.

There is one important missing element in determining the future of health care in Canada: the voice of patients and the public in general, both as citizens and as taxpayers. The only voice they have is through a general election. This is a major imbalance in the power structure, the dynamics of reforms, and the accountability mechanisms of the health care system.

1.1.1   SOME HISTORICAL FACTS

The Canadian Constitution of 1867 made HEALTH a provincial responsibility and this did not change with the 1982 repatriation of the Constitution from Britain. The Constitution confirmed the following direct responsibilities of the Federal Government: Veterans' and Natives' health care, drugs' administration and regulations with respect to drugs, devices, etc. The Constitution (1982), under its equalization provisions, obliges the provinces to provide comparable levels of public service for comparable levels of taxation. The first public discussion of a Canadian Health Care system occurred in 1919, on the platform of the Liberal Party of Canada. In 1945, there was a Federal-Provincial Reconstruction Conference. The Minister of Health of Canada, Brooke Claxton, proposed the National Health Grants Program, but this was opposed by Ontario (George Drew) and Quebec (Maurice Duplessis). In 1948, this was re-submitted by the new Minister of Health of Canada, Paul Martin, Sr. All provinces approved it and some funding was provided towards a comprehensive health insurance program.

Just prior to this time, T.C. (Tommy) Douglas, the Premier of Saskatchewan, passed *The Saskatchewan Hospital Services Plan* (1947). In 1957–58, the Federal Liberal government under Paul Martin, Sr., followed with an offer of a 50–50 share (Federal-Provincial); this was accepted by all as the *Hospital Insurance and Diagnostic Services Acts* (HIDS). In 1962, Saskatchewan was a pioneer again, despite a dramatic doctors' strike; it instituted *The Saskatchewan Medical Care Insurance Plan*. In 1964, Mr. Justice Emmett Hall's Commission set up by the opposition (Diefenbaker of the Conservative Party), reported to the Pearson Liberal Government, recommending a national Medicare program. In 1966, the Federal Government created *The Health Resource Fund* to help build hospitals and purchase equipment. In 1967, the federal Minister of Health, Allan McEachen (Liberal), succeeded in passing *The Medical Care Act*. By January 1971, all provinces had "Medicare" despite a bitter specialists' strike in Quebec in 1970.

The transfer of funds set-up by the *Canada Health and Social Transfer* (1996) replaced the old 50-50 formula and the provinces began to hold an increasingly large share of the costs of health care. However, the five fundamental conditions for health services across the country, as set up by the *Canada Health Act* (1984), were maintained. The five principles of the *Canada Health Act* are the cornerstone of the Canadian health care system and have iconic status for Canadians. This legislation, passed unanimously by Parliament in 1984, affirmed the Federal Government's commitment to a *universal, accessible, comprehensive, portable and publicly administered* health insurance system. The administration of the health care insurance plan of a province or territory must be carried out on a non-profit basis by a public authority. All medically necessary services provided by hospitals and doctors must be insured. All insured persons in the province or territory are entitled to public health insurance coverage on uniform terms and conditions. Coverage for insured services must be maintained when an insured person moves or travels within Canada or travels outside the country. Reasonable access by insured persons to medically necessary hospital and physician services must be unimpeded by financial or other barriers. *The Canada Health Act* contains provisions that ban extra-billing and user-charges, which means there is to be no extra-billing by medical practitioners or dentists for insured health services under the terms of the health care insurance plan of the province or territory; this also applies to user-charges for insured health services by hospitals or other providers under the terms of the health care insurance plan of the province or territory.

Although much of the Medicare costs are covered by public funds, hospital food, laundry and laboratory tests are frequently done in the private sector, therefore Canada currently has a mixed model rather than a social insurance one. The taxation base covers basic health care expenditures, which includes general taxation on personal income, on corporations, and on the sales of some goods and services. Charitable contributions cover some of the costs of equipment. The provincial government allocates a budget for hospital expenses related to continuing care, public health, mental health and rehabilitation services. A global budget is set for physicians' fees which are negotiated with the provincial medical association; the latter then allocates the funds by specialty.

Since 1980, the median for Western nations' health care expenditures as a percentage of the GDP was around 7-8%. In 1999, the median was 7.9%. That year, Canada was the fourth highest spender at 9.3%, equal with France. Below this benchmark were U.K., Spain, Finland and Japan. Countries with expenses higher than the median were Canada, France, Germany, Switzerland, and the U.S.. In 2011, Canada was at 11.2% and the U.S. was at 17.7% of GDP; the U.K. was at 9.4 and France at 11.6 (OECD, 2013_health care expenditures).

In 2009, Canada spent 8% from public funds and 3.3% from private ones; the U.S. spent 8.3% from public funds and 9.1% from private funds. The last ranking carried-out by the WHO on the quality of health care was done for the year 2000: Canada ranked 30th and the U.S. 37th (WHO). For the year 2008, life expectancy of women in Canada was 83.1 years, men 78.3, and mortality under the age of 5 was 5 per 1000. In the same year in the U.S., life expectancy for women was 80.6 years and for men 76, and child mortality was 8 per 1000. There are many other interesting statistics for all OECD countries on the website of that organization (OECD, 2013_life expectancy).

1.2   UNITED STATES

Defining the American health system, Holahan et al. (2007) wrote:

> In the U.S., direct government funding of health care is centered on Medicare, Medicaid, and the State Children's Health Insurance Program (SCHIP), which cover eligible senior citizens (65 years or over), the very poor, disabled persons, and children. The federal government also runs the Veterans' Administration that provides care to retired or disabled veterans, their families, and survivors through medical centers and clinics. The U.S. government also runs the military health system. One study estimates that about 25% of the uninsured in the U.S. are eligible for these programs but many people are not enrolled; however, extending coverage to all who are eligible remains a fiscal and political challenge (**Holahan et al., 2007**).

For everyone else, health insurance must be privately paid. In 2008, around 58% of U.S. residents had access to health care insurance through their employer (DeNavas-Walt et al., 2009) People whose employers do not offer health insurance and those who are self-employed or unemployed must purchase insurance on their own. People who enroll in Medicare need to buy Part A services and they can also buy Part B at additional cost. What Medicare does not cover for those enrolled in these plans are: Long-term care, routine dental or eye care, dentures, cosmetic surgery, acupuncture, hearing aids and exams for fitting them, and routine foot care. These can be covered by other insurance programs if provided by the employer or if purchased by the person (Medicare.gov, n.d.). The abstract from a paper by Ficella (2011) summarizes the state of health in the U.S. in that year:

> The United States has made little progress during the past decade in addressing health care disparities. Recent health care reforms offer a historic opportunity to create a more equitable health care system. Key elements of health care reform relevant to promoting equity include access, support for primary care, enhanced health information technology, new payment models, a national quality strategy informed by research, and federal requirements for health care disparity monitoring. With effective implementation, improved alignment of resources with patient needs, and most importantly, revitalization of primary care, these reforms could measurably improve equity (**Fiscella, 2011**).

Similar to the situation in Canada, the U.S. government is highly involved in health care through regulation and legislation. In 1973, the Health Maintenance Organization Act provided grants and loans to subsidize Health Maintenance Organizations (HMOs); the Act contained incentives to stimulate their popularity. HMOs had been declining before the passing of this law; but by 2002, there were 500 such plans enrolling 76 million people (Medical News Today, 2004).

1.3   COVERAGE AND ACCESS

In both Canada and the U.S., access to health care can be a problem. For many Canadians, the first point of contact for medical care is a doctor, so not having access to a regular medical doctor is associated with fewer visits to general practitioners or specialists, who can play a role in the early screening and treatment of medical conditions. In 2012, 14.9% of Canadians aged 12 and older, roughly 4.4 million people, reported not having a regular medical doctor. More males than females reported not having a regular medical doctor. However, when people cannot find a family doctor to enroll with, Canadian Citizens are still covered by the national health care system, so they can usually find a walk-in clinic and see the doctor on call (Statistics Canada, 2012).

In the U.S., a recent report by the Commonwealth Fund stated that the Affordable Care Act, which will go into effect in January 2014, will provide new insurance options for people without health insurance, enabling 14 million uninsured people to gain coverage in that year, and 27 million by 2021. The Commonwealth Fund Biennial Health Insurance Survey of 2012 found that half of adults (46%) aged 19–64 (estimated as 84 million people) did not have insurance for the full year or were underinsured and unprotected from high out-of-pocket costs; and 2/5 (41%) adults (75 million people) reported having problems paying their medical bills or were paying off medical debt; while more than 2/5 (43%) or 80 million people reported problems getting needed health care (Commonwealth Fund, 2013).

1.4   HEALTH CARE FACILITIES

The biomedical engineer and biomedical technician, to interface effectively with health care facilities (hospitals, clinics, etc.), should understand the structure of these organizations and the types of professionals with whom they are likely to interact. There are differences between teaching hospitals and non-teaching ones. The former are affiliated with a Medical Faculty in a University or with some of its medical programs and have a research role in addition to its clinical responsibilities. In Canada, non-teaching hospitals are community hospitals that do not have an academic tie. In the U.S., non-teaching hospitals are predominantly for-profit private institutions. In earlier studies, it was shown that teaching hospitals had a larger inventory of equipment and possibly more sophisticated and acute levels of care than community hospitals; their clinical engineering departments would have a larger budget per value of inventory than non-teaching hospitals (Frize, 1990b). Today, this is still the case in Canada, but not necessarily the case in the U.S.; see Wang et al. (2008) for more details.

Other factors that affect the role of a hospital and the size of its equipment inventory are: location (rural or urban), the acuity of patients treated, and the number of beds. Hospitals with tertiary services have the most acute beds (intensive care, cardiac care, etc.). Some medical specialties that may be found in tertiary level care hospitals are: cardiac surgery, neurosurgery, neonatal and pediatric intensive care, and other specialties that require the highest level of care. Hospitals with secondary-level care would have some operating rooms for less complicated surgeries. They would have medical imaging facilities and may offer dialysis and other medical treatments. Primary-level care hospitals would likely have out-patient clinics, and either short-term or longer-term patient care for chronic diseases; some would have some type of emergency services.

The health care professionals with whom clinical engineers and biomedical technicians interact most frequently are nurses. Nursing personnel are usually the people operating many of the medical equipment in the hospital. Examples are cardiac monitors, infusion pumps and controllers, defibrillators, pacemakers, incubators and many other types of devices. Clinical Engineering Departments (CEDs) also interact with respiratory and medical imaging technologists, audiologists, physiotherapists, and anyone else using medical devices in the health care environment. Physicians are important partners; they are most frequently in a position to request new equipment and to make choices on what they would like to have for making a diagnosis, a treatment, or to monitor patients.

The next chapter presents various sensor technologies to measure physiological variables in the human body in order to make a diagnosis, treatment, or monitor a patient's condition.

REFERENCES

Commonwealth Fund (2012). Report available at: http://www.commonwealthfund.org/Publications/Fund-Reports/2013/Apr/Insuring-the-Future.aspx; last accessed September 2013.

DeNavas-Walt, C, Proctor, BD, Smith, JC (2009). "Income, Poverty, and Health Insurance Coverage in the United States: 2008: Table 59." Available at: http://www.census.gov/prod/2009pubs/p60-236.pdf; last accessed July 2013.

Fiscella, K (2011). "Health Care Reform and Equity: Promise, Pitfalls, and Prescriptions." *Ann. Family Med.*, Jan, 9(1): 78–84. DOI: 10.1370/afm.1213.

Frize, M (1990b). "Results of an International Survey of Clinical Engineering Departments. Part II: Budgets, Staffing, Resources and Financial Strategies." *Med. Biol. Eng. Comput.*, 28: 160-165. DOI: 10.1007/BF02441772.

Health Canada (2010). "Health Care System." Available at: http://hc-sc.gc.ca/hcs-sss/medi-assur/index-eng.php; last accessed July 2013.

Holahan, J, Cook, A, Dubay, L (2007). "ORG1 Characteristics of the Uninsured: Who is Eligible for Public Coverage and Who Needs Help Affording Coverage?" Available at: http://kaiserfamilyfoundation.files.wordpress.com/2013/01/7613.pdf; last accessed July 2013.

Medical News Today (2004). "Health Care Expenditures in the USA." Available at: http://www.medicalnewstoday.com/releases/6225.php; last accessed July 2013.

Medicare.gov (n.d.). "What Medicare Covers." Available at: http://www.medicare.gov/what-medicare-covers/; last accessed July 2013.

OECD (Organization for Economic Co-operation and Development), 2013_health care expenditures, available at: http://www.oecd-ilibrary.org/social-issues-migration-health/total-expenditure-on-health_20758480-table1; last accessed September 2013.

OECD, 2013_Life expectancy, available at: http://www.oecd-ilibrary.org/social-issues-migration-health/life-expectancy-at-birth-total-population_20758480-table8; last accessed September 2013.

Statistics Canada (2012). Available at: http://www.statcan.gc.ca/pub/82-625-x/2013001/article/11832-eng.htm; last accessed September 2013.

Wang, B, Eliason, RW, Richards, SM, Hertzler, LW, Koenigshof, S (2008). "Clinical Engineering Benchmarking: An Analysis of American Acute Care Hospitals." *J. Clin. Eng.*, 33(1): 24-27. DOI: 10.1097/01.JCE.0000305843.32684.52.

World Health Organization (2000). "The World Health Report 2000 - Health Systems: Improving Performance." Available at: http://www.who.int/whr/2000/en/whr00_en.pdf; last accessed July 2013.

OTHER SUGGESTED READING

De la Maisonneuve, C, Oliveira Martins, J (2013) "Public Spending on Health and Long Term Care: A New Set of Projections." OECD report available at: http://www.oecd.org/eco/growth/Health%20FINAL.pdf; last accessed September 2013.

OECD health key tables (n.d.). Available at: http://www.oecd-ilibrary.org/social-issues-migration-health/health-key-tables-from-oecd_20758480;jsessionid=10da8h7ps450k.x-oecd-live-01; last accessed September 2013.

Wikipedia (2012). "Comparison of the health care systems in Canada and the United States." Available at: http://en.wikipedia.org/wiki/Comparison_of_the_health_care_systems_in_Canada_and_the_United_States; last accessed July 2013.

CHAPTER 2

Measuring Physiological Variables in Humans

There is a wide variety of technologies to measure human physiological variables and this is the subject of several books. This chapter provides an overview of common measurements performed on the human body and describes the types of sensors used to measure these physical quantities such as temperature, cardiac activity, blood pressure, chemistry values in the blood and urine, enzymes, and proteins.

Sensors are devices that provide an electrical signal when receiving a stimulus which could be a pressure, movement, chemical reaction, among others. A transducer is defined as a device that converts one physical quantity into another. "Sensors" and "transducers" are frequently used interchangeably. In this chapter, the word sensor is used for any device that allows the measurement of a physical quantity in humans. Testing in humans is a very important part of the medical act to establish a diagnosis for a patient's condition. Testing also allows caregivers to assess whether the treatment plan is improving the patient's condition and by how much. The major trend in biomedical sensor development is to design non-invasive testing mechanisms, which means they do not pierce the body at any point. However, there are still invasive techniques used regularly in medicine, as they may be more accurate; also, they are used where a non-invasive technique does not exist.

2.1 COMMON MEASUREMENTS OF PHYSIOLOGICAL VARIABLES

Readers should familiarize themselves with the physiology of the human body prior to learning about sensor technologies.

2.1.1 THE HEART

The heart has four chambers: two atria and two ventricles. There are two cycles in the heart: systole is the pumping phase in which the ventricles contract; and diastole is the resting phase and the filling of the atria. There are a number of measurements that can be done to assess the health of the heart and to identify diseased states. The electrocardiogram (ECG) is the electrical signal which can be captured with a set of electrodes placed on the chest, arms, and legs. Electrodes must make a good contact with the skin; the best ones are made of Ag-Ag-Cl (silver-silver-chloride) and contain an electrolyte as gel. Cardiologists are trained to read the signal recorded on paper to detect if the signal is normal or whether it contains abnormalities or irregular beats (also called arrhythmias); see Figure 2.1 for an example of the ECG lead connections and the signals they produce (Wikipedia_Cardiology). When monitoring the ECG of a patient for several days in a cardiac care center, the usual electrode configuration is lead-II, which consists of three electrodes. However, a 12-lead configuration is used when recording the ECG on a printer called an electrocardiograph, for diagnostic purposes. In addition to these measurements, the cardiologist can require a stress test that consists of a 12-lead ECG collected while the patient walks on a treadmill. This provides information when the patient is undergoing physical effort. Sometimes the physician orders a thallium stress test; in this case, the patient is injected with an isotope, then undergoes a gamma camera test (imaging of the heart), followed by a stress test, then a gamma camera imaging again. The gamma camera imaging shows the cardiologist if there are some areas of the heart that have been damaged, and the stress test may indicate whether some of the vessels near the heart have a blockage.

The ECG signals are quite small, usually of the order of 1 millivolt, so they must be amplified at least 1,000 times to be seen by instruments such as the cardiac monitor (similar to an oscilloscope) or on the electrocardiograph (printer). Another tool used by cardiologists is the Holter Monitor which records the ECG for a longer period (24 or 48 hours); the signal is played back in a rapid mode to assess the number and types of arrhythmias present during the period tested.

Common conditions that can be diagnosed with the electrical signal of the heart are: aneurysm, angina, angina pectoris, arrhythmias, atrial fibrillation, arteriosclerosis, bradycardia, congestive heart failure, a heart attack, tachycardia, and ventricular fibrillation.

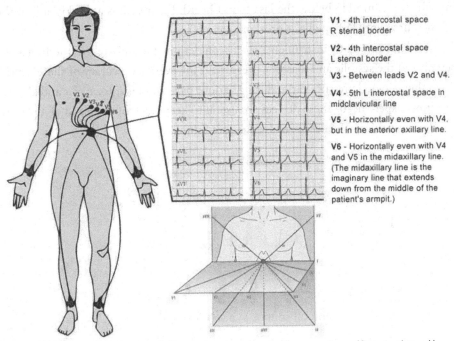

Figure 2.1: ECG lead connections and the respective signals they produce. (Source: http://www.ivline. info/2010/05/quick-guide-to-ecg.html)

Signal Conditioning for the ECG

In addition to the need for amplification, other important features are: the need to eliminate 50 or 60 Hz interference which can cause a wandering baseline; muscle tremor can also be a problem, so patients are required to lie down in a relaxed state for this test. There are filters in the amplifier to help reduce these problems.

2.1.2   NON-INVASIVE BLOOD PRESSURE MEASUREMENTS (NIBP)

The simplest way to measure arterial blood pressure is using a stethoscope and a cuff that can be inflated and deflated; this is called the Korotkoff sounds method. The cuff is inflated to a level above the systolic pressure and then the cuff is deflated slowly. Through the stethoscope, the nurse or doctor listens to the Korotkoff sounds; tapping, repetitive sounds for at least two consecutive beats occur at the systolic pressure point; the second sounds are the murmurs heard for most of the area between the systolic and diastolic pressures; the third sound is a loud, crisp tapping sound; the fourth sound, at pressures within 10 mmHg above the diastolic blood pressure, is described as "thumping" and "muting"; the fifth Korotkoff sound is silence as the cuff pressure drops below the diastolic blood pressure. The disappearance of sound is considered to be at the diastolic blood pressure point, which is 2 mmHg below the last sound heard. Figure 2.2 is a description of the signal shape and measurement (Wikipedia_Korotkoff_sounds).

Today, several automatic instruments have been designed to imitate this method and provide a numeric value of the systolic and diastolic pressures. However, the accuracy of these devices varies greatly. The value of blood pressure varies during various parts of the day or night and also according to levels of stress experienced. These devices are useful to keep track of the trend of a person's blood pressure and as a warning when abnormal values are found.

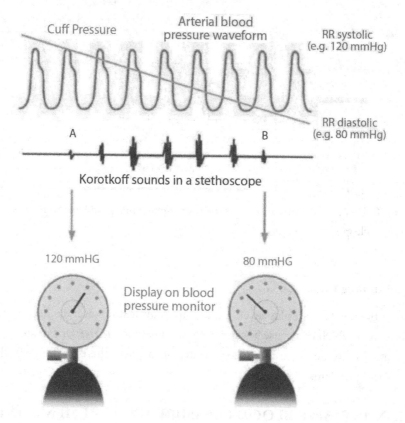

Figure 2.2: Korotkoff sounds to measure arterial blood pressure. (Source: adapted in English from: http://en.wikipedia.org/wiki/Korotkov_sounds)

2.1.3 INVASIVE BLOOD PRESSURE

For a more accurate measurement, especially when patients are in a critical or intensive care unit, invasive blood pressure measurement techniques are frequently used, especially for the long-term monitoring of cardiovascular instability. This allows to avoid the frequent use of an inflated cuff, thus preventing damage to the tissue of the arm; the invasive method is also reputed to be far more accurate for blood pressure measurement. A cannula made of Teflon or polyurethane is inserted into an artery, a 20G cannula for adults, but smaller (22G and 25G for children and neonates). Preferably, the radial or dorsalis pedis artery is used in adults to measure the arterial systolic and diastolic pressure (Ward and Langton, 2007).

For measuring central venous pressure, the catheter is inserted into a vein; this is similar to the pressure in the atrium of the heart. In some cases, the measurements are done directly into the chambers of the heart using a Swan-Ganz catheter to measure the right or left ventricular pressure and the pulmonary artery pressure; the pulmonary capillary wedge pressure is done by inflating the end of the catheter for a second or so. The catheter is connected to a three-way stop-cock and to a pressure transducer (sensor). A heparinized saline solution is connected to the stop-cock in order to flush the catheter periodically, which helps to prevent blood clotting at the tip of the catheter. Another of the outputs of the stop-cock is used to draw blood for blood gas testing, which is another important measurement for critically ill patients. The transducer is described in the physical sensors section further below. The saline bag is pressurized to about 200 mmHg to prevent the blood from backing up. The sensor must be placed at the level of the midline of the body measured at the chest area in order to produce accurate measurements. Invasive pressure measurements are only performed in a special unit (an intensive care or in a catheterization laboratory). See Figure 2.3.

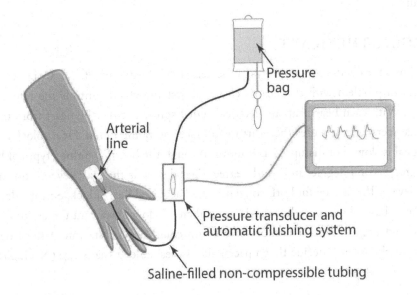

Figure 2.3: Blood pressure transducer and catheter. (Source: adapted from http://www.aic.cuhk.edu.hk/web8/haemodynamic%20monitoring%20intro.htm)

2.1.4 DEFIBRILLATION

When a person's heart stops or goes into fibrillation, it is desirable to perform cardio-pulmonary resuscitation and apply defibrillation through two paddles placed across the chest. Ventricular fibrillation is a condition in which there is uncoordinated contraction of the ventricles of the heart which quiver rather than contract, and so the pumping of blood stops. The paddles are either metal with gel applied for a good contact with the skin or they are disposable gel pads. The device provides an electric shock of short duration (around 5 ms) at a high voltage and current (typically up to 3000 volts at full power). This can help to bring the heart back into sinus rhythm (normal rythm). Automated external defibrillators (AEDs) are available in many public spaces such as airports, train stations, airplanes, ships, among others; they claim to be easy to use, especially if people have followed a cardiopulmonary resuscitation course (CPR) that includes training on how to use these defibrillators. They can save people's lives when used quickly on persons who are having a heart arrest. When the paddles are placed on the person, the device can identify whether a shock should be administered or not by identifying if the person's heart is in fibrillation or tachycardia (extra fast heart rate that can easily go into fibrillation). It is a simple process to charge the paddles with the high voltage and discharge them into the person. There are also small defibrillators implanted into the chest that discharge a voltage to the heart when the patient experiences a ventricular fibrillation or tachycardia.

2.1.5 PACING THE HEART

Figure 2.4 shows the conductive pathways in the heart (Wikipedia_Electrical_Conduction_Heart). Heart block is a problem with the heart's electrical system, which controls the rate and rhythm of heartbeats. With each heartbeat, an electrical signal spreads across the heart from the upper to the lower chambers, which causes the heart to contract and pump blood. Heart block occurs if the electrical signal is slowed or disrupted as it moves through the heart. The three types of heart block are: first degree, second degree, and third degree. First degree is the least severe, and third degree is the most severe. This is true for both congenital and acquired heart block. Second-degree Type I means the heart has a slower rate, but this is not sufficiently low to warrant the use of a pacemaker. However, second-degree Type II is more severe and could degenerate into a third degree heart block, so a cardiologist may decide that a pacemaker is desirable at that stage (NIH, 2012).

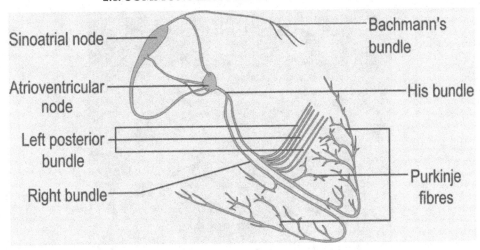

Figure 2.4: Pacing the heart. (Source: http://en.wikipedia.org/wiki/File:Conductionsystemofthe-heartwithouttheHeart.png)

Pacemakers can be used externally in an emergency situation. However, when this device is prescribed for permanent use, it is implanted into the chest. The pacemaker produces a signal that replaces the signal of the sino-atrial (SA) node when the latter does not send signals for the heart to pump blood (complete block) or only functions for part of the time (partial block).

There are several types of pacemakers. The single-chamber pacemaker has one pacing lead attached to a chamber of the heart, either the atrium or the ventricle. With the dual-chamber pacemaker, one lead paces the atrium and another paces the ventricle. This type closely resembles the natural pacing of the heart by coordinating the function of both the atria and the ventricles. The rate-responsive pacemaker is the most complex and expensive. Its sensors detect changes in the patient's physical activity and automatically adjust the pacing rate according to the body's needs. An image of a pacemaker and leads is shown in Figure 2.5.

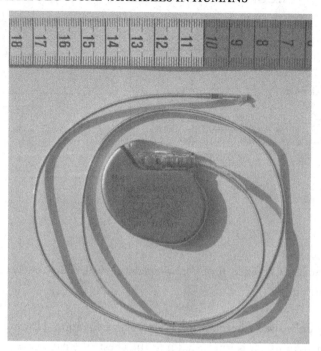

Figure 2.5: Typical pacemaker and leads. (Source: http://en.wikipedia.org/wiki/File:St_Jude_Medical_pacemaker_with_ruler.jpg)

2.1.6   THE BRAIN

The neurons in the brain emit signals, which can be recorded on the scalp; the electrical signal is called the electroencephalogram (EEG). It is used to diagnose epilepsy, tumors, stroke, coma, or brain death, among others. This signal is very small, of the order of 10 microvolts. The most common approach is to use small needle electrodes which are pinned on the scalp placed at specific positions. The international configuration is the most commonly used; it is shown in Figure 2.6. The 10-20 system refers to the distance between adjacent electrodes which are either 10% or 20% of the total front-back or right-left distance of the skull. Each site has a letter to identify the lobe and a number to identify the hemisphere location: F, T, C, P and O correspond to frontal, temporal, central, parietal, and occipital lobe, respectively. There is no central lobe; the "C" letter is only used for identification purposes. A "z" (zero) refers to an electrode placed on the midline. Even numbers (2, 4, 6, 8) refer to electrode positions on the right hemisphere, whereas odd numbers (1, 3, 5, 7) refer to those on the left hemisphere. Two anatomical landmarks are the points between the forehead and the nose and the lowest point of the skull from the back of the head, normally indicated by a prominent bump. (Wikipedia_electrodes_international system 2013)

Researchers in the 1930's and 1940's identified several different types of brain waves: delta waves (below 4 Hz) occur during sleep; theta waves (4-7 Hz) are also associated with sleep, deep relaxation and visualization; alpha waves (8-13 Hz) occur when we are relaxed and calm or during meditation; and beta waves (13-38 Hz) occur when we are actively thinking, problem-solving, or doing mathematical calculations. Today, more types have been added: the sensory motor rhythm (SMR, around 14 Hz) was originally discovered to prevent seizure activity in cats. SMR activity seems to link the brain and body functions. Gamma brain waves (39-100 Hz) are involved in higher mental activity and consolidation of information. (Brain and Health, n.d.)

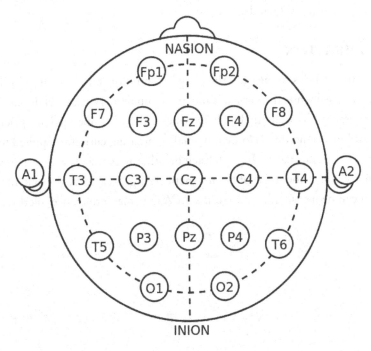

Figure 2.6: International electrode configuration for EEG. (Source: http://en.wikipedia.org/wiki/File:21_electrodes_of_International_10-20_system_for_EEG.svg)

2.1.7 ELECTROMYOGRAPHY (EMG)

Electromyography (EMG) is a test to assess the health of the muscles and of the nerves that control the muscles. A needle electrode is inserted through the skin into a muscle and the electrical activity is displayed on a monitor. Frequently, the electrical discharges are also heard with a speaker. The presence, size, and shape of the waveform, and the action potential produced on the monitor provide information about the ability of the muscle to respond when the nerves are stimulated. Nerve conduction velocity (NCV) is a test of the speed of conduction of impulses through a nerve; in this test, the nerve is stimulated, usually with surface electrodes placed on the skin over the nerve at various locations. One electrode stimulates the nerve with a very mild electrical impulse. The resulting electrical activity is recorded with the other electrodes. The distance between electrodes and the time it takes for electrical impulses to travel between electrodes are used to calculate the nerve conduction velocity.

The amplitude of this signal can vary between 20 and 2000 microvolts; the frequency is 6-30 per second. Pathologies can cause slower nerve conduction, less synchronization (scatter), lower amplitudes, and a modified shape of the signal. EMG signals have also been useful for many years to control artificial arms and hands.

2.1.8 RESPIRATION

There are various methods to measure the rate of respiration, but a common one, which is non-invasive and can be used while the person is moving (ambulatory), uses RIP bands (respiratory inductive plethysmography). Two elastic bands in which coils are embedded are placed, one around the chest and the other around the abdomen. A small alternating current is applied to the coils, creating a small magnetic field normal to the coil. Each breath causes a change in the cross-sectional area of the body, which changes the shape of the magnetic field generated by the belt, inducing an opposing current which can be measured, see Figure 2.7 (Wikipedia_inductance_plethysmography).

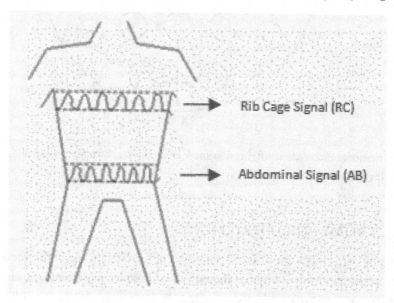

Figure 2.7: Rip bands to measure respiration cycle. (Source: http://en.wikipedia.org/wiki/Respiratory_inductance_plethysmography)

2.1.9   MEASUREMENT OF TEMPERATURE

Several sensors exist to measure temperature. The oldest method is the thermometer in which red-dyed alcohol is displaced in a glass tube that is calibrated to the temperature of humans. Electronic thermometers determine the temperature by measuring the change in electrical resistance of a metal wire or of a thermistor that is composed of a semiconducting material. The wire or thermistor is usually enclosed in a slender rod, or probe. A change in temperature causes a change in the resistance of the thermistor. The temperature is indicated on a meter or as a digital display.

An important use of thermistors is with an infant incubator. A sensor is placed on the infant's abdomen to measure the body temperature; this information is fed back to the incubator temperature control to maintain a constant ambient environment for premature babies who have an immature temperature regulation system. The thermistor chosen for this application is one that is linear for the small range of temperature needed to keep the infant warm; see Figure 2.8.

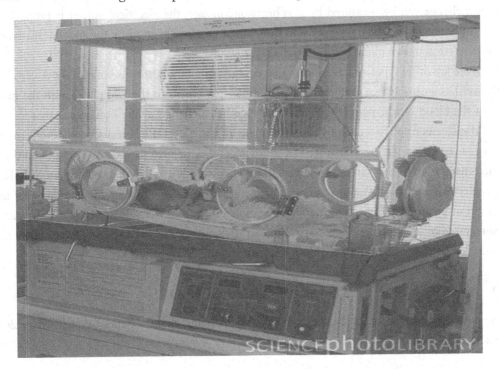

Figure 2.8: Incubator for premature infants. (Source: http://www.sciencephoto.com/image/290781/530wm/M8200156-Baby_in_incubator,_father_looking_on.-SPL.jpg)

2.2   TRANSDUCERS AND SENSORS

A sensor is a device that measures a physical quantity; there are many types of sensors and some examples are provided below. A transducer transforms one physical quantity that is captured into another form of signal, usually an electrical output. Both terms are used in the section below. Sensors, like transducers, can change the quantity measured into another form, which is why the two terms are often used interchangeably.

In physical sensors, the effect depends on the physical nature of the input stimulus; the output is usually an electrical signal. Examples are resistance sensors, inductive sensors, capacitive sensors, piezoelectric sensors, electromagnetic sensors, photoelectric sensors, and thermoelectric sensors (Wang & Liu, 2011). For example, a metal resistive strain gage is frequently used to measure invasive blood pressure. As blood exerts a pressure on a strain gage, it causes the elastic material to deform, which in turn causes a change in the resistance of the material of the strain gage; this resistance change can be measured by a Wheatstone bridge; the electrical output is proportional to the displacement of the strain gage and thus to the pressure exerted by the blood. This type of sensor is referred to as a blood pressure transducer.

There are other types of transducers that can measure blood pressure invasively. A capacitive transducer has a capacitance that changes with the displacement of the two plates with the movement of the blood on the plates. The inductive type has a change in the inductance that is based on the displacement of two coils by the movement of the blood. Piezoresistive materials such as silicon can be used in the same manner as a strain gage; they deform with displacement and the resistance change can be measured by a Wheatstone bridge. Today, with the development of MEMS (micro-electro-mechanical systems), pressure sensors (transducers) are very small and can be used at the tip of a catheter. They have many applications, including the measurement of intracranial pressure, intrauterine pressure in obstetrics, and air pressure during respiration to assess respiratory diseases. MEMS are also used in cochlear implants (Wang & Liu, 2011).

2.2.1   CHARACTERISTICS OF PHYSICAL SENSORS

The performance of sensors can be measured by the following characteristics (Wang & Liu, 2011; Webster, 2009).

The Transfer Function

The transfer function defines the relationship between the input quantity to be measured and the electrical signal produced at the output. This determines other parameters such as the **sensitivity**, which is simply defined as the amount of change produced in the output signal in response to a change in the input quantity measured. A large output signal for a small change in the input would represent a large sensitivity.

Span or Range

The span or range describes the range of input quantity that can be used with the sensor. For example, in a thermometer for human use, the ideal range needs to be close to a normal temperature, with a few degrees under and over the normal, representing the small range that can occur in live persons. Similarly, for blood pressure measurements, the range to measure arterial pressure (50–180 mm Hg) is larger than the measurement of venous pressure whose range is typically less than 20 mm Hg.

Hysteresis

The sensor output does not return to its original output value when the input quantity goes up and down or to a value of zero. Hysteresis is like memory and represents the size of the error in terms of the measured quantity. Sensors need to be calibrated regularly in order to function properly. For invasive blood pressure sensors, it is recommended to re-calibrate every time there is a shift of personnel (at 8 or 12 h) and more frequently if a problem in the output value is suspected. More frequent calibration is not justified because when the stop-cock is opened to the atmosphere for calibration infection can be a risk.

Non-Linearity

Non-lineraity refers to the maximum deviation from a linear transfer function. The error compares the actual transfer function to the best straight line.

Noise

All sensors produce some noise in addition to a signal. Normally, this is "white noise" which is composed of a signal with equal amplitude for a wide range of frequency.

Resolution

Resolution is defined as the smallest detectable input quantity change.

Bandwidth

All sensors have a finite response time to an instantaneous change in the input signal. Moreover, many sensors have decay times, which would represent the time after a step change in the input signal for the sensor output to decay to its original value. The reciprocal of these times corresponds to the upper and lower cutoff frequencies, respectively. The bandwidth of a sensor is the frequency range between these two frequencies

2.2.2   CHEMICAL SENSORS

Propertied by Properties

This type of sensor responds to a chemical reaction and produces an electrical signal proportional to the concentration of the biological analytes; the analyte can be one type or a multitude of types to be analyzed. The earliest chemical sensor was developed by Cremer in 1906; glass thin film responds to hydrogen ions in a solution, so Cremer developed a glass electrode to measure the pH. It is after the 1960s that chemical sensor development expanded rapidly.

Chemical sensors have been classified into three categories: ion sensors, gas sensors, and humidity sensors. Ion sensors are usually small, with a fast response, wide range, with high selectivity and sensitivity, and are not expensive. The principle behind an ion-selective sensor is as follows. The sensor has a membrane that separates the sample solution and a reference solution. Assuming that the membrane is only selective to one type of ions, then the diffusion occurring causes a potential to accumulate on the reference electrode (assume it is positive), then a negative potential accumulates on the ion-sensitive electrode. The difference of potential is proportional to the concentration of the specific ion in the sample solution (Wang & Liu, 2011). There are millions of chemical substances in the world and many types of sensors have been created to measure these. Some applications are in environmental, agricultural, weather, and health care measurements. In medicine, they include measuring the pH, sodium, and potassium in the blood. An example in health care is the measurement of glucose in the blood, which is important for diabetic patients; another is the measurement of urea in urine to assess the condition of the kidneys. Recent developments are the e-Nose, the e-Tongue, and the μTAS (micro total analysis system).

Electronic Nose

The **electronic nose** is a device designed to detect odors and flavors. There are possible applications in the field of health care such as the detection of bacteria; this is currently applied to recognizing the smell of the MRSA (Methicillin-resistant Staphylococcus Aureus), a highly resistant and dangerous bacteria. If such a system is placed in a hospital's ventilation system, it can detect and therefore prevent the contamination of other patients near those who are affected. Equipment could also be safer from being exposed to highly contagious pathogens. The e-nose can also detect lung cancer and certain other medical conditions by detecting the volatile organic compounds that are present with the medical condition. Nasal implants could warn of the presence of natural gas, for people who have a weak or no sense of smell (Wikipedia_electronic_nose).

Electronic Tongue

The **electronic tongue** is a device that can compare and measure different tastes. Current devices use seven sensors, each of which is capable of detecting dissolved organic and inorganic compounds that are normally detected by the human receptors. In fact, it is claimed that most of the detection thresholds of these sensors are similar or better than those of human receptors. In the biological mechanism, taste signals are translated into electric signals. The e-tongue sensors function in a similar way; they generate electric signals as potentiometric variations. Taste quality perception and recognition is based on the recognition of activated sensory nerve patterns by the brain and on the taste fingerprint of the product. This step is achieved by the e-tongue's statistical software which interprets the sensor data into taste patterns (Wikipedia_electronic_tongue). Some of the applications are: analyzing flavor aging in beverages; quantify bitterness or "spice level" of drinks or of dissolved compounds; quantify the taste masking efficiency of formulations such as tablets, syrups, powders, capsules, lozenges; and analyzing medicine stability in terms of taste.

Micro-Total Analysis System (μTAS)

The **Micro-Total Analysis System** (μTAS) is a laboratory on a chip (LOC). Because of the significant miniaturization, many tests can be conducted with a very small device built on a chip. These small devices provide low cost fast results of analyses at the point-of-care. Today, the devices can integrate several laboratory functions in a chip size of a few square millimeters to a few square centimeters in size. They need a small volume of the fluid to be tested, typically less than a pico liter (that is, 10^{-9}). The devices can range from simple channels to more complex gear wheels and levers, valves and pumps, all on a chip. They can be detectors like electrodes, fiber optics, and sensors; combining these can produce small complex instruments. Applications in the 1990s were in genomics, such as capillary electrophoresis and DNA microarrays. Because of their low cost, they may be appropriate tools for developing countries.

The most common material used is silicon and most fabrication processes use photolithography. New processes have been developed such as glass, ceramics and metal etching, deposition and bonding, and polydimethylsiloxane (PDMS) processing (Reyes et al., 2002). Some of the advantages of LOC are: low fluid volume, faster analysis, compactness of systems, and lower fabrication cost. Disadvantages are: this is a novel technology and therefore not yet fully developed; physical and chemical effects on a small scale may become more important and thus make processes more complex to deal with than conventional instruments; detection principles may not scale down well, leading to low signal-to-noise ratios; and although precision is high in micro fabrication, the geometric accuracy may be poor compared to precision engineering (Wikipedia_ MTAS).

2.3   GAS SENSORS

There are several types of gas sensors such as electrochemical, semiconductor, optical, solid electrolyte; most have industrial and domestic applications. For example, electrochemical gas sensors are used to detect low levels of toxic gases and oxygen in domestic and industrial environments. Important non-invasive sensors for medical applications are the transcutaneous partial pressure of oxygen in the blood (tc-pO2) and the partial pressure of carbon dioxide in the blood (tc-pCO2); pO2 and pCO2 are vital measures for babies on artificial ventilation; these transcutaneous measurements enable the medical team to reduce the amount of blood taken from the infant for blood gas analysis during the premature infant's stay in the neonatal intensive care.

It is possible to monitor continuously and non-invasively the pO2 of an infant using a heated Clark electrode on the skin. Changes in oxygen levels can cause significant complications in these infants such as retinopathy of prematurity (ROP) or broncho-pulmonary dysplasia (BPD). It is known that there are variations between transcutaneous values and those obtained by blood gas analysis using the infant's blood. But the monitors present a trend on a continuous basis, which is helpful for long-term monitoring. However, these values should be compared with the blood gas values at regular intervals. The other issue is that, for optimum operation, the Clark electrode must be heated to around 44° C. That can cause burns if the electrode is not moved every three to four hours. Considering that there may be two electrodes (O2 and CO2) on a very tiny body, this is not an easy approach, but it does help to limit the amount of blood loss for the infant. For the CO2 electrode, a study compared transcutaneous CO2 and end-tidal CO2 (ETCO2) to arterial measurement of CO2 and found that the tc-CO2 method was more accurate than the ETCO2 approach in ventilated pediatric patients who were four years old or more. The article mentions that the accuracy of tc-CO2 has been demonstrated in neonates because they have a thin skin with fewer diffusion barriers to capillary gases (Berkenbosch et al., 2001).

The next chapter discusses the equipment management process in health care institutions.

REFERENCES

Berkenbosch, J et al. (2001). "Noninvasive monitoring of carbon dioxide during mechanical ventilation in older children: end-tidal versus transcutaneous techniques." *Anesth Anal*, June, 92(6): 1427-31.

IVLINE (2000). "A quick guide to ECG." Available at: http://www.ivline.info/2010/05/quick-guide-to-ecg.html; last accessed July 2013.

NIH (2012). "Types of heart block." Available at: http://www.nhlbi.nih.gov/health/health-topics/topics/hb/types.html; last accessed July 2013.

Reyes, DR, Iossifidis, D, Auroux, P-A, Manz, A (2002). "Micro Total Analysis System. Introduction, Theory and Technology." *Anal. Chem.* 74: 2623-2636. DOI: /10.1021/ac0202435.

Wang, P. & Liu, Q (2011). "Biomedical Sensors and Measurement." In: s.l.:Zhejiang University Press, Hangzhou and Springer-Verlag, Berlin, Heidelberg: 51, 62, 119-121.

Ward, M, Langton, JA (2007). "Blood Pressure Measurement." *Cont. Ed. Anaesth. Crit. Care Pain.* 7(4):122-126. DOI:10.1093/bjaceaccp/mkm022.

Webster, JG, Ed. (2009). *Medical Instrumentation: Application and Design.* Wiley, 4th Edition.

Wikipedia, "Electrical_Conduction_Heart." Available at: http://en.wikipedia.org/wiki/Electrical_conduction_system_of_the_heart; last accessed July 2013.

Wikipedia, "Electrodes international 10-20 system for EEG signals." Available at: http://en.wikipedia.org/wiki/10-20_system_%28EEG%29; last accessed July 2013.

Wikipedia, (2012). "Electronic nose." Available at: en.wikipedia.org/wiki/Electronic_nose; last accessed July 2013.

Wikipedia, (2012). "Electronic tongue." Available at: http://en.wikipedia.org/wiki/Electronic_tongue; last accessed July 2013.

Wikipedia, "Repiratory inductance plethysmography." Available at: http://en.wikipedia.org/wiki/Respiratory_inductance_plethysmography; last accessed July 2013.

Wikipedia, "Micro total analysis system (lab-on-a- chip)." Available at http://en.wikipedia.org/wiki/%CE%9CTAS; last accessed July 2013.

OTHER SUGGESTED READING

Lee, SJ, Lee, SY (2004). "Micro total analysis systems (µ-TAS) in biotechnology." *Appl. Microbiol. Biotechnol.* 64: 289-299. DOI: 10.1007/s00253-003-1515-0.

CHAPTER 3

Management of Medical Technologies in Industrialized and Developing Countries

3.1 EVOLUTION OF CLINICAL ENGINEERING IN INDUSTRIALIZED COUNTRIES

The development of medical technologies began slowly in the first half of the 20th century. However, the 1950s and 1960s saw a major increase in the proliferation of devices applied to health care for making diagnoses, treatments, and to monitor patients. In this period, several departments of medical physics were created, whose main function was to calculate radiotherapy doses and to store and dispense isotopes for treatments. Biomedical technicians were hired in some of these departments to insure the maintenance and proper functioning of electronic and mechanical devices used in patient care.

A factor that encouraged hospitals to hire clinical engineers, at least in the U.S. and Canada, was an article by Ralph Nader in the *Ladies' Home Journal* in 1971 which claimed that there were five thousand deaths due to electrocution in U.S. hospitals every year (Nader, 1971). Nader's article was based on recent discussions by biomedical professionals regarding the possibility of electrocution when patients had a catheter lodged in the heart with external leads connected to some device; a leakage current that would be in contact with the external leads could result in a ventricular fibrillation and death if not reversed within a few minutes. Nader exaggerated the number of electrocutions per year which had been cited as 1,200 by Dr. Carl Walter in his article; Nader quoted several authoritative sources, but he hiked the number of deaths to 5,000 in his widely distributed article (Ridgway et al., 2004, p.9).

This adverse publicity spurred hospitals to hire engineers in the U.S. and in Canada, to ensure the electrical safety of patients. As a first priority, the clinical engineer and/or technician would test every medical device in the hospital for electrical safety. Problems found were fixed, but sometimes a recommendation of disposal had to be made. But it soon became clear that the majority of the work needed involved much more than electrical safety.

In the U.S. and Canada, there were two types of professionals hired: biomedical technicians and biomedical engineers. Some of these new professionals were given their own separate department and various names were used to define the new type of service: Clinical Engineering Department (CED); Medical Engineering; Biomedical Engineering. The name Medical Physics was used for a department with physicists, although sometimes these departments also had clinical engineers and/or biomedical technicians. In some cases, biomedical technicians and engineers reported to the Plant Operations Department which was normally responsible for electrical, mechanical and gas services. In the U.K., it was more common to see clinical engineers residing in a Medical Physics Unit. In France, several clinical engineers reported directly to a Medical Director (in charge of medical services) or to the Hospital Administrator. In the Nordic countries and in Europe, Clinical Engineering was frequently a part of a university-based biomedical program; it provided services to the university-affiliated hospitals and reported to the Department of Biomedical Engineering at the university.

In the 1980s, the field of Clinical Engineering expanded rapidly in several industrialized nations. Among the first to establish a medical equipment maintenance program in the U.S. was the Veterans Administration (VA); the VA had a national program by 1966. At the early stage of development of Clinical Engineering, hospitals focused on cost-effective repairs of equipment and they mostly hired technicians (Dyro, 1988; Dyro, 2004). However, this approach did not provide the engineering expertise which was needed in modern hospitals. When engineers were hired, attention became focused on the entire life cycle management of medical equipment. The clinical engineering department's role now encompasses electrical safety, pre-purchase consultation, incoming inspections, preventive and corrective maintenance, training of users on the safe and effective use of equipment, and de-commissioning of equipment when obsolescent. These activities are described in more detail further in this chapter (Frize, 1988). Clinical engineers and technicians support various specialists such as physicians, nurses, respiratory and X-ray technologists, and several other personnel in units where medical equipment is used.

In 1967, the American Association for the Advancement of Medical Instrumentation (AAMI) was created with the mission to support the healthcare community in the development, management, and use of safe and effective medical technology (AAMI, n.d.). AAMI develops many standards on medical devices with the collaboration of industry and professionals in the field. In 1970, AAMI began a certification process for biomedical technicians.

In 1971, The Emergency Care Research Institute (ECRI) was created. ECRI published *Health Devices*, a monthly publication containing reports of defective equipment or supplies as well as recalls for major hazards discovered with specific equipment or supply. For many decades and even today, the ECRI Institute has provided a comparative analysis of various medical devices, similarly to Consumer Report (ECRI, n.d.).

Another important step in the development of the discipline of Clinical Engineering was the establishment of a certification process to assess the competence of engineers and technicians. A few Canadian engineers and technicians including this author, received their certificate in 1973 from the U.S. Certification Board of AAMI. However, the certification process was later made available in Canada through the Canadian Medical and Biological Engineering Society (CMBES).

Another impetus to hire biomedical professionals arose from a guideline set in 1974-1975 by The Joint Commission (TJC), called Joint Commission on Hospital Accreditation (JCAH) at the time), that hospitals establish a medical equipment safety program. The accreditation is to allow hospitals to participate in the Medicare-Medicaid programs and receive reimbursement for services rendered. It is voluntary, not mandated by government, but if licensing is required by the states, then it becomes mandatory. TJC visits hospitals and provides an accreditation standing to those that meet the requirements on all aspects of health care delivery as per the TJC guidelines. In Canada, in view of the increased use of medical devices in patient care and of the move by the Joint Commission on Accreditation of Hospitals in the United States to include aspects of medical technology management as part of their survey of hospitals, the Canadian Council on Hospital Accreditation asked the CMBES to prepare a brief on the proper role of clinical engineering in Canadian hospitals. The brief prepared by the CMBES outlined seven basic principles associated with clinical engineering, hospital accreditation, and the proper management of medical technology in hospitals. These principles helped to bring some changes to the Canadian Hospital Accreditation Guide and Questionnaires that began to include some aspects of technology management (McEwen, 1982).

In 1976, the *Journal of Clinical Engineering* was established. In the same year, the U.S. Congress enacted the Medical Devices Amendments to the Food, Drug, and Cosmetic Act enabling the Food and Drug Administration (FDA) to regulate the medical device industry (Dyro, 1988).

From a global perspective, the International Federation of Medical and Biological Engineering (IFMBE) set-up a Working Group for Clinical Engineering in 1979 which held its first meeting in Marseilles in 1980. Original members were: A. Oberg (Sweden, Chair), C. Roberts (U.K.), D. Kraft (GDR), Z. Katona (Hungary), and M. Frize (Canada); the mandate was to help develop the field of clinical engineering in the 33 countries where biomedical societies were affiliated to the IFMBE. The IFMBE, through its Working Group, published a set of criteria that represented the minimum qualifications that clinical engineers should have to be recognized by their profession through certification processes to be held in various countries. In 1985, the Working Group was transformed into the Clinical Engineering Division (CED), which still exists today. Early work consisted of organizing and co-sponsoring workshops and seminars to discuss the role and place of clinical engineering within health care institutions. Workshops were held in London (U.K.) in 1986, Trondheim (Norway) in 1987, San Antonio (U.S.) in 1988, Patras (Greece) in 1989, Weimar (GDR) in 1990, Kyoto (Japan) in 1991, and Fontevrault (France) in 1992. The discussions led to a clearer definition of the role of this profession within health care institutions and encouraged several groups of clinical engineers in various countries to share information and expertise. This helped the Division to develop and publish a guideline in 1991 entitled *Mutual Recognition of Qualifications for Clinical Engineers.* Recent activities and reports of the Clinical Engineering Division can be found on the website of IFMBE (ifmbe.org).

One of the early definitions of the role of a Clinical Engineering Department (CED) was published in an article in 1988: "A CED should provide safe and effective management of technology used in patient diagnosis or therapy, within health care institutions" (Frize, 1988). In 1991, the American College of Clinical Engineering (ACCE) was founded, an organization committed to enhancing and supporting the profession of clinical engineering. The website of ACCE posts a definition of the clinical engineer: "A Clinical Engineer is a professional who supports and advances patient care by applying engineering and managerial skills to healthcare technology" (ACCE, n.d.).

Another factor that helped the new field of Clinical Engineering to develop was the rapid proliferation of medical technologies in the 1980s. For example, in a community hospital of 540 beds in New Brunswick, Canada, the CED was responsible for the management of 320 devices in 1979, and managed more than 2000 devices by 1989, an increase of 600%. The CED at the hospital provided reports to the hospital administrators showing that in-house equipment management delivered substantially more comprehensive services for one-third of the cost of services provided by external companies (Frize, 1989). This type of feedback to decision-makers ensured substantial support and recognition of the critical role played by clinical engineers and biomedical technicians in the process of health care delivery.

A 1988 survey of CEDs in western nations, including the U.S and Canada, Sweden and Finland, with a few respondents in the European Community, found that there were some differences in the level of involvement of engineers and technicians in various aspects of the medical device management functions. But overall, the field developed in a fairly similar way in most of these geographical regions. The majority of CEDs were involved in pre-purchase consultation, drawing specifications and requirements for device acquisitions, analysis of quotations and issued recommendations based on criteria established with the users (physicians, nurses, and other health care staff). The CED performed corrective and preventive maintenance and incoming inspections when equipment was delivered; several were involved in training users on the safe and effective use of devices. Some of these functions were performed by biomedical technicians and others by clinical engineers. Unfortunately, the study found that half of the CEDs responding to the survey did not think that their work was receiving recognition from their hospital and some were not consulted prior to equipment purchases in their institution. This study led to the development of a model describing a desirable level of involvement in technology management, and defined the resources needed to accomplish these tasks (Frize, 1990a,b). A longitudinal survey carried out in 1991 found that things had not changed much in the three year span between the previous survey (1988) and the new one (Frize, 1994).

In 1999, another survey based in large part on the Frize 1991 survey, involved six regions: North America, Nordic countries, Western Europe, Southern Europe, Australia, Brazil, Mexico and

Cuba. Results were similar to the previous surveys discussed above, except that a slightly higher proportion of CEDs felt recognized for their work (Glouhova, et al., 2000). Later, in 2007, a new survey of CEDs was carried-out in developing countries. Although this was also based on the 1988 Frize survey, the new survey added questions regarding equipment donations (Mullally and Frize 2008; Mullally 2008). One conclusion was that the Frize survey questionnaire was still applicable in 2007 to both developed and developing countries. The newer study enabled to assess the resources needed by CEDs in developing countries and added new and interesting aspects about donations of equipment to these nations.

The education level of the engineers and technicians in industrialized countries differs somewhat from those hired in developing countries. In the 1991 Frize survey, half of the departments were staffed by engineers with a Master's degree in biomedical engineering; just under 25% of engineers had a doctorate; 13% had a Bachelor's of Engineering; and 10% of the departments did not have an engineer. For the technicians, 60% had completed a two-year program, and 23% a four-year program. A few had done a three-year program and only 2% had one year of post-secondary education. In developing countries, the education levels were somewhat lower; details are provided in a later section (Frize 1994; Mullally and Frize 2008).

As mentioned previously, certification of clinical engineers (CCE) and of biomedical equipment technicians (CBET) began in 1973 with AAMI, both for the U.S. and Canada. The CCE process is currently managed by the Healthcare Technology Foundation (HTF) and the biomedical technicians by AAMI (HTF n.d.; AAMI_certification).

3.2 FUNCTIONS AND ACTIVITIES OF CLINICAL ENGINEERING DEPARTMENTS (CEDS)

3.2.1 IN-HOUSE MEDICAL EQUIPMENT MAINTENANCE

Most CEDs perform this role and concentrate their activities on corrective maintenance and preventive maintenance of medical equipment; this includes equipment used for monitoring patients, treating them, or making a diagnosis; it excludes medical imaging and laboratory equipment. Some Clinical Engineering departments have added medical imaging and laboratory equipment to their workload, but it is likely that the most complex equipment such as the Magnetic Resonance Imaging (MRI), Computer Tomography Scanner (CT-Scan), or large and complex chemistry analyzers are covered by a service contract with an external company, either with the Original Equipment Manufacturer (OEM) or with a third-party service agency. Some of the equipment is highly complex, but others are more easily tackled, especially if some training has been provided to the biomedical technicians and if they have good service manuals. The other category of equipment sometimes managed by CEDs includes anesthetic machines and artificial ventilators and respirators.

Another service function is the incoming inspection, which is usually performed by the technicians. This consists of verifying the safety and performance of equipment delivered to the health care facility. Incoming inspections are done for new devices just purchased or a device repaired by an outside agency and returned to the owner. In both cases, it is critical to ensure that the device operates within its specifications, that all accessories and parts are delivered as well as operating and service manuals; most Canadian hospitals require that equipment be certified by the Canadian Standards Association (CSA); in the US, National Fire Protection Association (NFPA) and the Underwriters Laboratory (UL) have standards that can be adopted by authorities such as hospitals, which then makes them mandatory (UL; NFPA).

3.2.2   FUNCTIONS USUALLY PERFORMED BY ENGINEERS

Equipment Acquisition

Whenever new equipment has been approved for purchase, the clinical engineer, in collaboration with the department responsible for purchases, must identify all potential users of the new technology and meet with them in a timely manner to discuss what criteria are needed and where the equipment is to be used. The acquisition is either a completely new technology or a replacement of existing equipment. The clinical needs define the functionalities and characteristics of the technology and the clinical engineer can then define the desired specifications; this document describes the clinical functions that the device is expected to provide, the accuracy of the measurements or of the output, the list of accessories necessary to operate it, the safety requirements, and any other information which will enable to compare various devices when companies bid on a request for proposals (RFP). Cost is one factor, but should not be foremost in the decision except for an equal performance on all points specified. Sometimes it is important to mention the requirement for compatibility with other devices that are currently used in the medical facility where the new technology is to be used. Today, requirements would also frequently specify that the output signal or data collected be exported and stored for future analysis and research. Ensuring that proper operating and service manuals are included is critical and the specifications should request service training if it is needed by the CED staff to maintain the equipment after the warranty expires. All quotations (bids) must be assessed against the criteria and specifications established for the acquisition. The final decision on what equipment to purchase should be made by the users in collaboration with the CED and the purchasing department.

Analysis of Alerts, Potential Hazards, and Incidents Possibly Caused by Medical Equipment

The ECRI Institute issues information called Alerts Tracker on a regular basis to their members. Equipment manufacturers are also expected to issue alerts or warnings when a problem is discovered. ECRI provides equipment evaluations in a similar way to the Consumer Reports; this can be useful to consider when acquiring new equipment. A main source for assessing hazards is to consult the U.S. Food & Drug Administration (FDA) website for recalls of devices and equipment. There are three levels of recalls: a Class I recall is the most serious, when there is a probability that the use or exposure to a product will cause serious adverse health consequences or death; a Class II recall refers to a situation where exposure to a product may cause temporary or medically reversible adverse health consequences or where the probability of serious adverse health consequences is remote; a Class III recall is a situation where exposure to a product is not likely to cause adverse health consequences. Clinical engineers should screen alerts and recalls received and check the inventory of equipment owned by their health care institution to check if any of these were identified by the alert or recall. A solution has to be applied in a timely fashion depending on the seriousness of the problem. The manufacturer usually provides what is needed to fix the problem or the device may have to be sent back to the manufacturer for an update.

Investigation of Incidents or Accidents

Whenever an accident or incident occurs in a hospital, it is important to assess whether a medical device was involved or not. There may be several reasons other than a defective device for an incident to occur such as use error, or a combination of circumstances that leads to a problem. However, medical equipment can be used correctly and still cause a problem if it develops a fault while being used. The CED should be involved in the process of identifying what happened. One method of analyzing an incident or accident is to do a simulation, trying to reproduce as closely as possible the circumstances that led to the problem. Human factors engineering is a useful approach to perform this type of study (Easty et al. 2009).

When there are incidents or accidents, a report must explain the potential causes and include recommendations to avoid future mishaps. In cases where a serious injury or death occurs, it is possible that some institutions require a report from a clinical engineer from a different organization. The lawyer hired by the institution where the incident occurred also needs to be consulted about what should be included in the report. Whenever there is an incident or accident that could have been caused by medical equipment, the clinical engineer should investigate the potential cause in collaboration with others responsible for this type of investigation.

Training Users on the Safe and Effective Use of Medical Technologies

CEDs should ensure that users receive adequate training from the manufacturer or the distributor of new equipment acquired. This would normally be done in collaboration with the department head of the unit concerned. In addition to this, the CED should provide workshops on the safe and effective use of equipment, on a regular basis, for all staff working with medical equipment.

This could be arranged with groups and especially include new staff hired during the course of the year. The frequency depends on the staffing level of the CED and on the needs of the health care facility. Note that it is not recommended that CE staff operate the equipment themselves as they are not qualified to perform clinical functions.

Management Functions

The Director of the CED should include in his/her duties the following: ensure that the staff has the resources needed to accomplish their tasks; prepare and submit the budget that allows the department to carry-out its functions and activities; evaluate the staff at regular intervals and provide feedback on how they can improve their work. In addition, to meet the criteria of the Accreditation of Health Care Facilities, the CED must develop and implement a Continuous Quality Improvement (CQI) program focused on key functions for which the department is responsible. This is required by the Accreditation Agencies both in Canada and the U.S. There is a difference between efficacy and effectiveness. A simple way to explain the difference between the two is that the latter refers to "doing the right things" whereas the former means "doing things right."

It is important for CEDs to measure their productivity and cost-effectiveness. However, this is quite challenging as there are many ways to assess these and each approach has advantages and limitations. For an interesting discussion on this topic, see Wang et al., 2012. Measuring the productivity of the CED staff is a measure of efficiency. The traditional way to measure productivity was proposed in several articles in the 1980s and has been a topic of discussion for some 30 years. Some of the models proposed to assess productivity were:

1. productivity (%) = time worked/time available X 100 (ASHE, 1982);

2. productivity (%) = chargeable hours/worked hours X 100 (Furst, 1987);

3. productivity (%) = earned hours/worked hours X 100 (Bauld, 1987); and

4. productivity (%) = hours worked/available hours X 100 (Frize, 1989), which is similar to the model suggested by ASHE (1982).

See a detailed discussion on this topic that presents various manners to assess productivity and issues related to each choice (Wang et al., 2012).

In her study of productivity, Frize (1989; 1990a,b) found that hospital size was not very useful as a criterion to determine staffing levels of CEDs; this is explained by the fact that hospitals differ in the number of acute care beds and thus in the amount of equipment needed to provide care. Equipment inventory size and replacement value were found to be more useful quantitative parameters to define workload related to the technology management functions.

Although any of the proposed productivity measurement approaches should not be used alone, as the issue is more complex than indicated by the suggested measurements, it is still useful for a CED to use a simple approach to assess in a general way if the staffing level is sufficient to carry out the workload for which it is responsible. One way to do this is to calculate the number of working days in a year (this excludes holidays, weekends, average vacation days per staff person, and average sick leave days taken by the staff.) The remaining days are multiplied by the hours of work per day, minus breaks and lunch. This number should be multiplied by a realistic percentage that takes into account interruptions, making reports of work done, etc. The final number can be interpreted as follows: a percentage between 75% and 85% of the total available hours suggests that the staff is logging in much of their work. Any number higher than 85% would be questionable as it is not possible to have 100% of recorded productivity. If the technicians justify between 60 and 74% of their time on assigned activities, this is considered acceptable, while recording a percentage between 50 and 59% indicates a need to find improved efficiencies for some of the tasks, or an easier way to record the time spent of their tasks; a productivity measurement under 50% is unacceptable. It is important to use this information with care, as if it is perceived as a close monitoring of the staff's use of time instead of useful statistics to populate the department, resentment can set in and the information provided may become falsified.

If the recording of the hours spent of various tasks are fairly accurate, then the department workload can be calculated by the average hours per year of corrective maintenance, preventive maintenance, incoming inspections, etc.. This number is then divided by the number of full-time-equivalent staff. This could enable the CED to assess if it has enough staff to do all the expected work or to justify hiring more people if it can show that it is understaffed for its level of activities.

3.3   A CHANGING ROLE IN THE 21ST CENTURY

The role of CEDs began to change in the past two decades. Most medical devices incorporate some form of computer technology, even for simple thermometers. Increasingly, medical equipment includes a feature that allows data collection from patients. A new role for CEDS is to help ensure that the data collection, storage and retrieval is available for clinicians and researchers. This new systems aspect of the role means that clinical engineers need to work more closely in collaboration with the hospital's Information Systems (IS) department who had, until recently, been responsible for the support of all computer systems such as the administration and payroll systems, a computerized pharmacy system, a patient information system (PIS), and several others as they were acquired (Zambuto, 2004). Grimes suggests that clinical engineers must act as the "stewards" of health care technology. As effective stewards, clinical engineers need to understand the practitioners' intent, they need to possess knowledge with respect to both existing and developing healthcare technology, and they must understand the various implications of applying the technology (Grimes, 2004).

In the past, the IS department and the CED have not really worked closely together. However, in some hospitals, the CED is responsible for IS staff and the opposite occurs in other sites. The meshing is not always smooth, but it is the most logical way to ensure an effective transition for hospitals to a computerized environment for all its functions. The Electronic Medical Record (EMR) is becoming a common feature and is discussed in Chapter 8 of Part II. Another new aspect of the CED role is to help reduce medical errors and adverse events through the use of appropriate information technologies. More details are provided in Chapter 7 of Part II.

Finally, it is a good practice for the staff of a CED department to make rounds and visit each clinical department, collect feedback on the perception they have of the equipment maintenance and management service provided, and query whether there are other problems that should be addressed. Many issues and problems can be identified with these rounds or visits and even by stopping in the corridor or in the cafeteria to chat for a few minutes with users. They often say: "I was going to call you about this…" Health care personnel are very busy with patient care and equipment issues are not foremost in their mind. So it is a good principle to seek interactions. When all goes well, this is also good news for the CED, and people come to understand the important role it plays in the health care team.

3.4   CLINICAL ENGINEERING IN DEVELOPING COUNTRIES

In developing countries, clinical engineering appeared later than in Western countries; this development occurred mainly in the late 1980s and early 1990s, except in countries like India and Brazil where they sprung up in the late 1970s and early 1980s. An international survey was conducted in 2003 to assess the status of Clinical Engineering Services delivered to hospitals in developing countries (Cao 2004). Data was collected from Asia, Africa, Latin America, and Mexico. The responses were compared to two previous studies done in industrialized countries (Frize 1989; Glouhova 2000). In addition to this study, a model of medical technology acquisition and diffusion was developed by Roy (2004). Another study was done in 2007-2008 in developing countries with a much larger sampling than the former study of 2003 (reported in 2004 by Cao) (Mullally 2008). The results of these three studies pertaining to developing countries are summarized below.

The 2003 questionnaire, available in three languages (English, French, and Chinese) using close-ended questions and boxes for many of the responses was sent to 700 contact persons in three regions: Asia, Africa, and Latin America. Sixty-four responses were received. Because of the small size of the sample, the responses were grouped into two regions: 34 from Asia (India, Bangladesh, P.R. China, Indonesia, Saudi Arabia, South Africa; and 27 from Latin America (Venezuela, Brazil, and included Mexico). The proportion of respondents from teaching hospitals was 65% (22/34) in Asia and 44% (12/27) in Latin America; 80% of departments existed as a separate unit, similar to the previous Frize survey in industrialized countries which was 81%. More than 80% of respondents were satisfied with their reporting authority. Like other regions studied by Frize (1990; 1994) and Glouhova (2000), respondents who reported to a "Senior administrator" were more satisfied than those reporting to a "Plant Director" or a "Medical Director." Respondents reporting to "other directors" were also satisfied; this category included "Lab Director," "University Technology Research Institute," "Logistic Department," and "Nursing Manager" (Cao 2004; Frize et al., 2005).

The education level and staff ratio of clinical engineers and technicians varied in the two regions. In Latin America, staffing was similar to industrialized countries, where technicians made up the majority of the staff. Clinical Engineers had at least a BSc, but a few had a MSc. or a Ph.D. In previous studies in industrialized countries, just over 60% of clinical engineers had a MSc. or a Ph.D., compared to 6% in Latin America and 4% in Asia. In Asia, the education level of clinical engineers was lower; 12% did not have a university degree, and fewer had an MSc. or Ph.D. degree than in Latin America. A few CEDs in Asia employed more clinical engineers than technicians, and a third (10/34) employed only technicians; 12% of CEDs in Latin America hired only clinical engineers (Frize et al., 2005).

The education level of technicians in Latin America had a slightly lower education level than Asia. The highest education in Asia was 3-year technical school after high school, compared to 2-year technical school in Latin America. But both regions showed a lower level of education of technicians than industrialized countries, where a number of technicians had a 4-year technical education. Only 26% (16/61) of respondents were members of professional or technical associations; 90% of CEDs in both regions trained their staff at special centers and/or on the job. Some respondents reported that manufacturers and dealers provided some of the training (Frize et al., 2005).

The number of devices and their replacement value managed by CEDs in developing countries were substantially lower than in industrialized countries. Repairs, incoming inspections, preventive maintenance, user training, and pre-purchase consultations were all activities performed by CEDs. However, research activities were not common. This differed somewhat from industrialized countries.

In terms of workloads, clinical engineers in this survey spent 40% of their time on equipment repairs, while those in Frize's survey spent the same amount of time on consulting activities. Technicians in the present study spent more time on repairs, incoming inspections than those in Frize's study, and less on preventive maintenance; their main responsibility was equipment repair. In Asia, clinical engineers performed more repairs than technicians; perhaps their qualifications were more like those of technicians in industrialized countries. This was not observed in Latin America. There were other responsibilities reported such as "bio-safety," "administrative tasks," "collaborative project," "tracing suppliers." "Administrative tasks" were frequent in Latin America with a workload ranging from 15%–68%; perhaps this coincides with the low hiring of clerical staff (Frize et al., 2005).

There were 30% (18/61) of CEDs stating that they did not have adequate manuals for equipment management. Compared with Latin America, Asia had a higher need of the manuals: 41%

(14/34) vs. 15% (4/27). Some of the reasons given by respondents were: "need more specific manuals," "need electric circuit diagrams," "need manuals in our local language," and "incoming devices without manual." On the question regarding the adequacy of their test equipment, spare parts, and space allocation per person, respondents in both regions differed from responses in industrialized countries; most respondents stated having inadequate resources, except for space allocation in Asia; while most respondents in the two developed country studies thought they had adequate resources. In the present survey, CEDs in Asia were satisfied with their space allocation when they had 15 square meters per person, while CEDs in Frize's survey were satisfied only with 20 square meters per person.

In this survey, all respondents who used computers for their equipment management were in Latin America, whereas all respondents who used manual records were in Asia. One third of CEDs using computers developed the software themselves. Almost all (98%) respondents in this survey stated having at least one computer in the department, and 96% (26/27) of Latin American respondents said they could always access the Internet from their departments. However in Asia, only 47% (16/34) had Internet access and all respondents stated not being able to go online. Quality assurance was carried-out by 70% of respondents in both regions; 80% of respondents in Asia had begun to perform productivity assessments, while 37% of Latin American respondents had begun to do this. In the survey, only 44% (20/45) of respondents felt recognized in their hospitals, which was similar to the earlier Frize study (1990). However, in the later Glouhova study (2000), more than 80% of respondents felt they were recognized.

3.4.1   A MODEL TO ASSESS THE READINESS OF A COUNTRY TO ACQUIRE, DIFFUSE, MANAGE MEDICAL TECHNOLOGIES

In her study, Roy examined e-commerce and e-government models, but none had been applied to a medical equipment situation. Roy developed a preliminary model to assess a developing country's readiness to acquire, diffuse, and manage medical technologies (Roy, 2004). Two countries with a vast difference in their level of development were used to construct the model. The model's attributes were extracted from data collected on Mali and Brazil. The data came from both primary and secondary sources, obtained mainly from CIDA (The Canadian International Development Agency). Data was collected on:

1. the principal health care concerns;

2. government spending in health care per capita;

3. whether the country had a strategic plan and/or policy to manage its health care system and equipment acquisition and management;

4. whether the country regulated the acquisition, use, and diffusion of medical devices;

5. whether there were internet and telephone services available to the population and to health care facilities;

6. organizational capabilities to manage the technology; and

7. health technology diffusion, and the level of technology available in the country's health care system.

Attributes that contributed to substantial health-related advancement were considered as enablers and the opposite were constraints. The seven categories of attributes were studied for each of the two countries and the results provided a risk rating of high, medium or low with regards to the potential failure or success of funding health-related projects in these countries. This research filled an urgent need for agencies like CIDA to evaluate the risk of potential investments in health-related projects. Roy's study also identified areas that Mali and Brazil should address to improve their medical technology readiness assessment. In the case of Brazil, connectivity for rural areas would support further deployment of medical technologies. In the case of Mali, the need was pressing in all qualifiers/enablers. The areas requiring improvement could become the object of future donor programming in the country (Roy, 2004). Decision-makers can pay particular attention to the enablers that scored high if they relate to the implementation of an initiative. For example, when considering an initiative related to medical staff certification in Mali, it is important to consider that the regulatory function is weak. The team should then include components that address those identified as high risks in their design.

The application of the medical technology assessment model to Brazil and Mali was encouraging. It provided a simple framework, recommending the avoidance of certain technologies because the country was not ready to adopt and sustain them. An interesting observation from this work was the trade-off between complexity and benefit; medical technologies stood to have the best health benefit at the primary care level. Paradoxically, that is where implementation and sustaining effort needed are more difficult. Another observation was that there probably exists a correlation between a country's medical technology readiness and its focus on health. An assumption would be that countries focusing on curative health are less likely to absorb and sustain new medical technologies than the ones which include a preventive focus (Roy, 2004).

3.4.2   TECHNOLOGY MANAGEMENT

Returning to the issue of technology management, although the number of respondents from developing countries in Cao's survey was not very large, the results were useful to assess where countries stood in terms of resources, responsibilities, and recognition. The study enabled CEDs in all developing countries to define their workloads and thus they could request appropriate resources. Roy's model to assess readiness of countries to acquire and diffuse medical technologies helps to identify weaknesses and the level of equipment sophistication that a country can sustain success-fully. Both studies can help developing countries to strengthen all aspects of medical technology management needed to improve health care delivery to their populations (Frize et al, 2005; Roy, 2004; Cao, 2004).

A new study in 2007-2008 collected 169 responses from 43 countries on clinical engineering effectiveness in hospitals. The responses were primarily from Africa, Latin America and Asia, with some representation from the Middle East and Eastern Europe. The data identified hospital and clinical engineering department profiles, human and equipment resources, equipment procurement and donation processes, with a focus on the role of the clinical engineering department (Mullally, 2008). Some of the findings were:

- Teaching hospitals accounted for 27.2% of overall responses; the most common hospital type was publically funded (40.2%) and 22.5% were private; philanthropic/non-govern-mental organizational hospitals were 9.5%. In Africa, the hospital size was 50-200 beds, in Latin America, they were 50-250 beds with some larger ones 250-500 beds; in Asia, the majority of hospitals had over 500 beds; in the Middle East and Eastern Europe, most had 250-500 beds with a few over 500 beds (Mullally and Frize, 2008).

- Only 57% of CEDs surveyed in this study existed as a separate unit, which is much lower than found in industrialized countries. A minority of departments (36% or 45/125) shared maintenance services with other hospitals. The majority of these (66.7%) reported being the main center of service.

- Regarding staff levels, all regions claimed not having adequate staff to carry-out their workload. When asked if CEDs had difficulty in finding qualified engineers and tech-nicians, all responded yes. For the difficulty of finding engineers, 79% of respondents from Africa and Latin America said yes; in Asia, the proportion was lower: 56.5%. In the Middle East and Eastern Europe, this was the highest at 83.3%. For finding qual-ified technicians, 70% of respondents in Africa and 77.6% from Latin America needed to find staff with these qualifications; in the remaining regions, the proportion was 60% (Mullally and Frize, 2008).

- A strong determinant of CED effectiveness is the level of involvement in the equipment acquisition process. 80% of participants reported the existence of an official procurement policy at their hospital, and almost half were "very" involved in the process. Regarding who led the procurement of equipment process, and who was involved in the team, see Table 3.1.

Table 3.1: Proportion of type of personnel involved in leading and in the decision-making process for equipment acquisitions (Source: Mullally 2008)

Personnel	Leads the process	Involved in decision
Administrator	45.7%	53.3%
User	27.6%	71.4%
Consultant	8.6%	36.2%
CED manager	30.5%	67.6%
CED staff	10.5%	53.3%

The majority of equipment was procured through a formal acquisition process (median of 90%), while donations accounted for 5%, leases, rentals, and loans amounted to another 5%. Donated equipment often arrived without appropriate resources. The proportion of respondents who claimed that spare parts never accompanied donations (n=63) was 58.7%; for user manuals, 29.7%; for maintenance manuals, 42.2%; for user training, 50.2%; and maintenance training, 61.3%. Often, there was little to no consultation with recipient hospitals when equipment was donated. The level of consultation with the hospital before a donation occurred was rated by respondents as "fair"; however, 36.2% reported that there was no consultation at all.

These studies have been helpful in assessing the level development of clinical engineering in both industrialized and in developing countries; it also enabled to identify their level of integration into the health care team and the resources they are given to perform their role. The data can help CEDs to develop a plan to move to the next level of involvement and to apply for the resources needed to get there.

3.4.3   WHO BASELINE COUNTRY SURVEY ON MEDICAL DEVICES (2010)

In 2010, the WHO carried-out a survey in 196 countries, receiving 144 responses. The survey results were reported at the First WHO Global Forum on Medical Devices held in Thailand in 2010 (WHO, 2011). The survey asked whether countries had: a national policy for health technology; guidelines for donations; technical specifications for procurement or donations; a national list of approved medical devices; a medical equipment management unit; and availability of high cost medical devices. For more details, see the entire report which also contains 42 recommendations for future actions (WHO, 2011).

The next chapter discusses safety considerations, minimizing liability, and continuous quality improvement program design.

REFERENCES

AAMI (n.d.). "Association for the Advancement of Medical Instrumentation." Available at: www.aami.org/; last accessed July 2013.

AAMI_Certification. Available at: http://www.aami.org/certification/about.html; last accessed September 2013.

ACCE (n.d.). "American College of Clinical Engineering." Available at: http://www.accenet.org/; last accessed July 2013.

American Society for Hospital Engineering (ASHE) (1982). "Part II: Determining Productivity, in Medical Equipment Management in Hospitals." Edition. Chicago, II, ASHE: 2-7.

Bauld, TJ (1987). "Productivity: standard terminology and definition." *J. Clin. Eng.*, 12: 139-145. DOI: 10.1097/00004669-198703000-00014.

Cao, X (2004). "Assessment of Clinical Engineering Department in Developing Countries." MASc Thesis in Systems Science, University of Ottawa.

CMBES (n.d.). "Clinical Engineering Standards and Practice." Published by the Canadian Medical and Biological Engineering Society. Available at: http://www.cmbes.ca/index.php?option=com_content&view=article&id=80&Itemid=99; last accessed July 2013.

Dyro, JF (1988). "Clinical Engineering: A Prospectus." s.l., *IEEE*: 80-85.

Dyro, JF (2004). *Clinical Engineering Handbook.* Ed. Elsevier Academic Press.

Easty, AC, Cafazzo, JA, Chagpar, A (2009). "Improving safety in healthcare through the establishment of a healthcare human factors team." In *World Congress on Medical Physics and Biomedical Engineering*, September 7-12, Munich, Germany, Springer Berlin Heidelberg: 324-327. DOI:10.1007/9783642038853_90.

ECRI Institute,(n.d.). https://www.ecri.org/Pages/default.aspx; last accessed July 2013.

FDA (n.d.). "U.S. Food and Drug Administration." Available at: www.fda.gov; last accessed July 2013.

Frize, M (1988). "The Clinical Engineer: A Full Member of the Health Care Team?" *Med. Biol. Eng. Comput.* 26: 461-165. DOI: 10.1007/BF02441912.

Frize, M (1989). "Evaluating the effectiveness of clinical engineering departments in Canadian hospitals." Doctoral dissertation, Erasmus Universiteit, Rotterdam, The Netherlands.

Frize, M (1990-a). "Results of an international survey of clinical engineering departments Part 1." *Med. Biol. Eng. Comput.*, 28: 153-159. DOI: 10.1007/BF02441771.

Frize, M (1990-b). "Results of an international survey of clinical engineering departments Part II." *Med. Biol. Eng. Comput.*, 28: 160-165. DOI: 10.1007/BF02441772.

Frize, M (1994). "Longitudinal study of clinical engineering departments in industrialised countries (1988 to 1991)." *Med. Biol. Eng. Comput.*, 132: 331-335.

Frize, M, Cao, X, Roy, I (2005). "Survey of Clinical Engineering in Developing Countries and Model for Technology Acquisition and Diffusion." Proc. IEEE/EMBC Shanghai: 171-173. DOI:10.1109/IEMBS.2005.1616369.

Furst, E (1987). "Guest Editorial—special issue: productivity & cost-effectiveness." *J. Clin. Eng.*, 12: 99-100.

Glouhova, M, Kolitsi, Z, Pallikarakis, N (2000). "International survey on the practice of clinical engineering: mission, structure, personnel, and resources." *J. Clin. Eng.*, 25(5): 205-212. DOI: 10.1097/00004669-200025050-00009.

Grimes, SL (2004). "Clinical engineers: stewards of healthcare technologies." *Eng. Med. Biol. Mag. IEEE*, May-June, 23(3): 56-58. DOI: 10.1109/MEMB.2004.1317982.

HTF (n.d.). "The Healthcare Technology Foundation." Available at: http://www.thehtf.org/; last accessed July 2013.

IFMBE (n.d.). "Clinical Engineering Division." Available at: http://ifmbe.org/organisation-structure/divisions/clinical-engineering-division/; last accessed July 2013.

McEwen, JA (1982). "Clinical engineering and hospital accreditation in Canada." *J. Clin Eng.*, 7(1): 83-85. DOI: 10.1097/00004669-198201000-00011.

Mullally, S (2008). "Clinical Engineering Effectiveness in Developing World Hospitals." MASc Thesis in Electrical Engineering, Systems and Computer Engineering, Carleton University, Ottawa.

Mullally, S Frize, M (2008). "Survey of Clinical Engineering Effectiveness in Developing World Hospitals: Equipment resources, procurement, and donations." Proc. IEEE/EMBC, Vancouver: 4499-4502. DOI: 10.1109/IEMBS.2008.4650212.

Nader, R (1971). "Ralph Nader's Most Shocking Exposé." *Ladies' Home J.*, 3: 98-179. DOI 10.1007/BF02368486.

NFPA (n.d.). "National Fire Protection Association." Available at: www.nfpa.org; last visited September 2013.

Ridgway, MG, Johnston, GI, McClain, JP (2004). "History of Engineering and Technology in Health Care." 9-10. In Dyro, JF (2004) *Clinical Engineering Handbook*. Elsevier Academic Press.

Roy, I (2004). "Medical Technology Assessment Model for Developing Countries." MASc Thesis in Telecommunications Technology Management, Carleton University.

UL (n.d.). "Underwriters Laboratory." Available at: www.ul.com; last visited September 2013.

Wang B, Rui T, Fedele J, Balar S, Alba T, Hertzler LW, Poplin B (2012). "Clinical Engineering Productivity and Staffing Revisited. How Should It Be Measured and Used? J. Clin. Eng., October/December: 135-145.

World Health Organization (2011). "First WHO Global Forum on Medical Devices: context, outcomes, and future actions. Available at: http://whqlibdoc.who.int/hq/2011/WHO_HSS_EHT_DIM_11.11_eng.pdf; last accessed September 2013.

Zambuto, RP (2004). "Clinical Engineers in the 21st Century." *IEEE Eng. In Med. & Biol. Magazine*, 23(3): 27-31. DOI:10.1109/MEMB.2004.1317980.

OTHER SUGGESTED READING

Dyro, JF (2004). *Clinical Engineering Handbook*. Ed. Elsevier Academic Press. 674 pages.

Miesen, M (2013) "The inadequacy of donating medical devices to Africa." *The Atlantic*, Sept 20. Available at: http://www.theatlantic.com/international/archive/2013/09/the-inadequacy-of-donating-medical-devices-to-africa/279855/; last accessed September 22, 2013.

Wang, B (2012). *Medical Equipment Maintenance: Management and Oversight*. Morgan & Claypool, Synthesis Lectures on Biomedical Engineering.

CHAPTER 4

Safety Considerations, Minimizing Liability, and Continuous Quality Improvement (CQI)

In this chapter, we examine the safety of patients and staff in health care institutions, more specifically electrical safety and hazards created by electromagnetic interference (EMI). The question of liability is discussed as well as how to minimize exposure. The last part of the chapter provides a short overview of quality assurance leading to a continuous quality improvement (CQI) program for Clinical Engineering Departments.

4.1   ELECTRICAL SAFETY IN HOSPITALS

Each country has its own standards and regulations regarding electrical safety in hospitals and other health care institutions. In Canada, the Canadian Standards Association (CSA) developed standards and regulations, not only concerning the distribution of electricity in the building, but also concerning the level of leakage current permissible on equipment found in patient areas, whether these are devices used in patient care or not. Hospitals in Canada are obligated to have all electrical supply and devices meet CSA requirements and all equipment entering the hospital must be CSA approved or tested.

In the United States, there are several laws, regulations, and codes that apply to the safety, maintenance, and management of medical technologies. The government body entrusted with the responsibility to regulate medical devices and drugs is the U.S. Food & Drug Administration (FDA). The FDA's primary task is to review the safety and efficacy of drugs and devices prior to their introduction into the market. Manufacturers are required to perform registration, labeling, and good manufacturing practices and to report adverse effects. The FDA is also involved with performance standards and pre-market approvals.

The Association for the Advancement of Medical Instrumentation (AAMI), in collaboration with The American National Standard Institute (ANSI), developed the standard: ANSI/AAMI ES60601-1:2005, *Medical Electrical Equipment—Part 1: General Requirements for Basic Safety and Essential Performance*. The third edition of the standard covers any medical device that requires an electrical outlet or a battery (AAMI).

The National Fire Protection Agency published *NFPA 99 – Standards for Health Care Facilities* (NFPA, 2012). Although it is a voluntary standard, NFPA 99 has been adopted by many states, counties, and municipalities through reference in their licensing requirements, thus becoming mandatory requirements for hospitals, but not necessarily for the manufacturers and other organizations located outside of health care facilities.

The *2012 NFPA 99: Health Care Facilities Code and Handbook Set* addresses the safety of patients, staff, equipment, and facilities; it covers a wide range of systems within healthcare such as electrical, gas and vacuum, environmental, materials, electrical equipment, and gas equipment; it also discusses other hazards such as fire and explosion, burns, chemical, and radio-frequency interference. The section on electrical equipment specifies certain design and construction criteria, as well as periodic tests and inspections, and contains requirements for user and service manuals, service documentation, and qualification and training of users and maintainers (NFPA.org).

4.2   BRIEF SUMMARY OF ELECTRICAL SAFETY ISSUES

Every device powered by electricity has leakage current that originates from capacitive, inductive, or resistive sources and can be present on the chassis or accessible metal parts of the device. This is not necessarily due to a fault and is normally channeled safely by a third wire connected to the ground. However, if the ground wire is broken, the leakage current on the metal parts of the device can be channelled through a patient connection and cause an electrical shock; if the level of this current is high due to a short or a fault, this can be dangerous for patients and staff, especially if there is a connection to the heart. There can also be a voltage difference between two devices connected to a patient, which may cause a leakage current to flow from one to the other through the patient.

A macroshock is an electrical current which flows on the outside parts of the body; this can be dangerous if the current level is sufficiently high and if it passes through the region of the heart, as between an arm and a leg. The body response to various levels of current has been documented by a few studies: A tingling sensation can be caused by a current of 0.5–2 mA; a 20 mA current causes a "no-let-go" situation, which means that the person cannot disconnect from the source. At 50 mA, there can be respiratory distress; 100 mA can lead to ventricular fibrillation, which can cause death if not reversed within a few minutes (Roy, 1980).

A microshock is a current that is applied to an internal part of the body near the heart. This can be sent through a catheter lodged near the heart, with metal contacts outside of the body, as with temporary pacemaker leads. In this case, the current level of 100 µA is reported to be able to cause a ventricular fibrillation. The body is more sensitive to currents and voltages that are in the frequency range of 50–60 Hz. Other common examples of potential internal connections are: intra-cardiac ECG leads, a fluid-filled line into an artery or a vein, or a pulmonary artery catheter such as a Swan-Ganz line connected to a pressure transducer, and drug/dye injections. For more details on the effect of current on the human body, see Roy, 1980.

4.2.1   ENSURING ELECTRICAL SAFETY

The best way to ensure that all devices are safe to use with patients and staff rests with proper design and manufacturing. However, it is also important to establish a schedule for safety tests on medical equipment in case a fault has developed or if the ground wire has been damaged over time. This entails measuring the leakage current and assessing whether it meets the various standards and regulations; it also tests if the ground wire in in good condition, that is, has an almost zero resistance. If the level of current is found to be higher than the acceptable limits for both microshocks and macroshocks, depending on the intended use and the presence of an internal connection or not, then a solution must be found. An isolation transformer can be added to limit the leakage current on the patient leads and chassis; another approach is to ensure all metal parts that can carry current are covered with an insulating material. If the equipment is very old, it could be decommissioned. Optic coupling for monitors is now a common way to isolate the patient from the electrical power circuit of the device.

3.2.2   ELECTRO-MAGNETIC INTERFERENCE (EMI) AND MEDICAL DEVICES

This section presents a brief historical review of the issues and discusses the EMI environment, that is, sources of EMI that can interfere with medical devices. Examples are provided of medical devices that can malfunction when in the presence of EMI. Some solutions are suggested to protect the devices from EMI and patients from potentially hazardous malfunctions.

The EMI Environment

Sources of EMI that can be present in a health care environment, or near patients who are connected to medical devices or who have devices implanted in their body, are presented below.

Sources of EMI

There are many sources of interference, such as local high power AM/FM and TV transmitters, paging systems, cellular telephone base stations and repeaters, two-way radios, amateur or CB radio, wireless communication devices, microwave ovens, and static discharge from humans. In the operating theater, there are electrosurgical generators and electric drills; in physiotherapy departments, we find diathermy, ultrasound therapy, and interferential therapy machines. All of these sources produce EMI that can be measured in Volts per meter. It is obvious that the closer the patient is to the source, the higher will be the emitted signal. The signal strength (usually in volts per meter) is divided by the square of the distance from the source. Medical data transmitted by wireless telemetry is another potential source of interference and is also vulnerable itself to interference from other devices. The transmission frequency of telemetry systems is usually selected in a way to minimize these problems.

Potential Impact of EMI on Medical Devices

The best way to demonstrate this potential danger is through examples. Take a patient with an implanted pacemaker of the "demand" type. This means that the pacemaker assesses the time between heart signals (ECG), and if this is longer than expected, it provides an impulse to the SA node and the heart then goes through its normal cycle. However, if there is an EMI signal, this can resemble a heart signal and so the pacemaker can remain silent. One can see that with continuous EMI signals present and an SA block, where the patient's heart does not pace by itself, then the patient can die. The same situation applies to an apnea monitor that measures the time between breaths; if there is an apnea (no breathing) and EMI signals are present, the apnea monitor can be silent and not sound an alarm. Again, if this lasts for a few minutes, the patient can die. An implanted defibrillator is also another device that can stop functioning properly in the presence of EMI.

Other types of medical devices can malfunction in a different manner. For example, patient monitors can lose their signal and stop monitoring the patient. This is a problem in the operating theatre when drills or electrosurgical generators are used. The anesthetist needs to have the signals on the monitor to assess how the patient is doing under the anesthetic agent and gases being administered. A good practice is for the surgeon to use the electrosurgical generator in spurts of 30 s or less; in this way, the monitor can regain its signals between the use of cutting and coagulation functions of the electrosurgical generator.

Infusion devices can be a serious problem as they can be reprogrammed to open flow by some EMI signals. If the drug being administered is potent, this can kill the patient or do some severe damage that may or may not be reversible. Artificial ventilators can also be reprogrammed and change the breath rate, which can cause severe problems for patients. Pulse oximeters may have false readings and incubators and radiant warmers can stop working or be reprogrammed. Some types of hearing aids can send loud signals to patients wearing them. It is also known that electrically powered wheel chairs can veer off course suddenly when in the presence of an interference signal; they can experience incidents of uncontrolled movement or electromechanical brake release that can lead to serious injury or death (Witters 2009).

It is important to note that EMI issues change with time and environment. Some hospitals or patients using medical devices at home are located in areas with transmitting sources, while others are in areas with no or low level sources. It is expensive to monitor EMI and it is frequently intermittent, so it may be quite difficult to capture. Devices can also behave differently in a laboratory when we try to reproduce the situation. Another issue is that physiological and biological signals are typically very low (in the order of mV or μV). We must use high gain amplifiers to be able to observe the signal on an output device. Leads, cables, and the human body can act as an antenna that captures the interference signals. There is also the phenomenon of the coupling of signals and rectification signals at electrode-skin interfaces.

Users (patients and the health care team) need to be aware of potential problems caused by EMI; they must question whether this is present when a malfunction occurs and report the event to the appropriate authority (especially to biomedical staff if they exist in the health care facility) to ensure others become aware of the problem detected.

Electro-Magnetic Compatibility (EMC)

EMC is the opposite of EMI. It refers to the level of EMI that a device can be subjected to without its proper functioning being disrupted or altered. In general, medical devices are expected to have an EMC of 3 V/m for both non-life and life support devices (IEC standard, 2004).

When problems are observed in a certain location with a specific device, say an EMG (electromyograph) signal collection and recording instrument, there is a solution: move the EMG device to a more EMI quiet location or remove the source of EMI. For example, in the 1970s, in a hospital in Montreal, the EMG equipment was reproducing radio signals instead of the patient's signals. On the other hand, the ECG department had very little or no EMI in their location. The two departments (ECG ad EMG) agreed to switch their location and the problem was solved. ECG signals do not experience EMI issues to the same extent as EMG signals. The hospital was located near major television transmission antennas. Some people suggested installing a Faraday cage, but I disagreed as this is a very expensive proposition and difficult to implement properly with just one single ground and no ground loops. A better solution is to carefully select the location of devices where the biological signals are very small. Tests can be done to see if the EMI signals are low for such clinics. Testing equipment is very expensive but perhaps can be borrowed for the tests when deciding to do renovations or when moving clinics to other locations. My solution in the 1970s and 1980s, since I could not easily find test equipment to borrow, was to connect a coil to my oscilloscope and see where interference signals were picked up and whether they were low or high. This was a crude approach, but it worked in the case of relocating the EMG clinic.

Classification of EMI incidents

Researchers use the term true positive when they can replicate an incident at a site and/or in a laboratory, which increases the degree of confidence on the nature of the particular incident. A false positive is an incident that is attributed to EMI, but is probably not due to EMI; it is perhaps a software problem or another type of malfunction. A false negative is a situation where an incident is not attributed to EMI, but it is likely occurring because of EMI; this often happens if the person reporting the incident is not familiar with the concept of EMI; sometimes this is subsequently confirmed by an alert from the manufacturer, or from organizations like the FDA in the U.S. or Medical Device Bureau in Canada.

As we saw previously, EMI can cause a variety of problems: devices operating outside of their specifications; operator intervention may be needed; or a fault may need a repair. The device's operation can be out of control; there can be a silent malfunction (no alarm or not functioning at all); or there can be a discrepancy between a clinical reality and the readout of the device. At other times, it may be impossible to read the signals, as in the example of the EMG machine discussed above. Moreover, some devices are unreasonably susceptible, so this should be checked when buying new devices. The FDA website can be consulted for a description of many EMI incidents that were reported to that agency (FDA.org).

4.3   LIABILITY EXPOSURE

This section discusses the liability exposure to malpractice suits in the health care field for clinical engineers and technicians. Approaches are suggested to minimize exposure. One question is whether personal malpractice liability insurance is needed. In general, it may not be necessary to buy this unless someone is an independent consultant. Hospitals should usually cover their employees under their own liability insurance policy. It is possible for the engineer to be sued individually for personal negligence, but this type of insurance is very expensive and not really a practical approach for a clinical engineer's salary range. But if engineers are in private practice, then it may be advisable to obtain coverage.

4.3.1   HOW TO MINIMIZE RISK

One way to minimize the risk of exposure is to ensure that the CED's program of activities is well planned, implemented, up-to-date, and meets all legal requirements in their region or country. It is especially important to keep documentation on the development of the department, of each intervention made by staff, and of each incident reported; the reports should include recommendations on how to avoid the recurrence of accidents or incidents.

The type of documentation that should be provided to the management team concerns the appropriateness (or inappropriateness) of purchases made by the hospital, compliance with codes and regulations, safety policies and tests, and records of the education and training of the department's personnel. Documents should mention what technical documents are available; describe the equipment management program, and what tests are done to limit exposure to liability; other documents that should be included are: reports of maintenance and repair costs; departmental performance (staff, budget, etc.); departmental objectives; departmental growth; and up-to-date information on potential hazards.

4.3.2 LEGAL ISSUES IN CLINICAL ENGINEERING PRACTICE

If a medical device causes an incident or accident, it is important to assess whether the product fails to meet reasonable expectations of the consumers (users) and, as a result, was the proximate cause of an injury. This can refer to an item, a piece of equipment, a device, the supply of services, or even a building. It can also refer to a procedure or a system.

A defect usually refers to a defective design or a failure to warn of certain hazards. A defective design can be in the specifications or blueprints, or appear in the composition, the process of manufacturing, testing, or inspection. Questions are: Did the device meet the standard of design and development? Is it safe for its intended use, with reasonably anticipated considerations? Is a safe or better design feasible? Does the device (design) create an unreasonable risk of harm?

Another aspect is the failure to warn; here we should consider the likelihood of an accident and the degree of danger created when it is used without a warning. We should also consider the feasibility of adopting a more effective warning. However, there is no duty to warn if the danger is obvious to the user such as "a knife is sharp." There is also no duty to warn against the abnormal use of the device.

Nevertheless, if there is an obvious danger or an unusual use is foreseeable, the courts may say there was a duty to warn; this depends on the utility, or severity of potential injury, the cost or ease of providing the warning, and the level of reduction of the danger with the warning. If the danger is not known, it may depend on the scientific knowledge of the time, whether a suspicion of danger exists, to decide on the practicality of adding an effective warning. If a warning is warranted, another consideration is the extent to which the warning must be communicated to potential users.

If there is a duty to warn, to whom should the warning be addressed. Aspects to consider are: the physical adequacy of the warning, its prominence, clarity, completeness, and method of transmission. Compliance with statutory, regulatory, and/or industry-wide standards does not necessarily avoid liability exposure. The duty to warn is a continuous process; it depends on progress, development and availability of new information; designers must alert users of discoveries either in the form of product notes, newsletters, letters to users, through persons in their sales force, or with labels; all of these approaches must be kept up-to-date.

For product liability, negligence can be avoided by exercising due care in design, manufacturing, distribution, testing, and marketing of devices. The most frequently used avenue is the case of strict liability. For example, if a failure is material in nature, the person responsible for this is held liable to the individual injured, as a result of a negligent conduct.

In engineering, due care means a careful evaluation of the concept, plan, materials used, process of manufacturing, testing, inspection, and the need for safety measures. If unavoidable danger exists, the duty to warn or to provide direction on its use is expected. The concept of negligence refers to the absence of due care or the failure to exercise due care.

In cases of strict liability, the injured party must prove that damage resulted from a condition of the product; or a condition was unreasonably dangerous; or an unreasonably dangerous condition existed at the time it left the defendant's control. It is not required that a fault be established in order to justify a recovery under strict liability. Misrepresenting a material fact about a product is answerable in damages through negligence and strict liability.

A tort is a legal wrong committed upon a person or property independent of any contract; this applies to negligence and strict liability.

A case of a breach of warranty is contractual in nature and is said to be the least frequently chosen in court cases. A warranty is an implied or expressed promise.

A legal defense would attack the basic elements of the proof required by the injured party, which normally would establish a defect or a duty to warn, or the need for a safety device. The amount of recovery may or may not be affected by the injured party's conduct at the time of the incident.

Contributory negligence is a failure to exercise due care to discover defects or to guard against their existence. This is of great importance for clinical engineers; their role is to minimize hazards and to screen alerts.

Product liability is of importance to manufacturers and distributors of equipment. Information on this topic is available from the FDA and from an article by Keeton (1972). Also, a detailed description of product liability can be found in the legal dictionary by Hill and Hill (1981-2005).

4.4   QUALITY ASSURANCE AND CQI

Quality assurance (QA) has been an integral part of hospital services for several decades and all departments are expected to develop a continuous quality improvement (CQI) program. When this requirement was added to guidelines for hospital accreditation, the term used was Quality Assurance (QA), which measures how well a function is being carried-out.

QA applies to services, not to products; it implies a measure of the level of service provided by a department or a group and a method to identify if additional efforts would improve the level of service and its quality. It must provide a guarantee that the level of excellence resulting from the QA program produces an acceptable level of patient care.

Clemenhagen (1985) provided suggestions on how to make such programs work. The author stated that an important ingredient for a successful implementation was a strong interest and commitment from the staff to be involved and to develop a program that applies to their particular service. Other advice was that staff must understand and believe in QA and focus on solving problems that are relevant and useful. A strong coordinating body is critical to ensure the integration of all the diverse programs into a whole that is relevant and appropriate for the hospital and its patients (Clemenhagen 1985).

4.4.1 HOW TO MEASURE QAA

After having identified a situation, a service, or an activity that can be improved, a comparison is done between actual results recorded and pre-established performance criteria for that activity or service. If it is not achievable, then poor morale will ensue. The goal is to increase the quality of the service; it should not be too easy to reach, otherwise the QA becomes meaningless. It is important that the criteria be realistic and achievable, and that it can be done with the resources available.

4.4.2 GENERAL GOAL

The general goal is to increase the effectiveness, efficacy, and quality of services to ensure the delivery of good patient care. The hospital's mission should form an integral part of a QA program. For clinical engineering departments, specific goals should be:

- ensure equipment is available and working within its specifications and safety standards for patient care;

- identify additional services needed which are not provided at this time;

- improve the efficacy and competence of the technical staff;

- minimize risks for patients and staff through training staff on the safe and effective use of medical devices;

- improve the comfort of patients if possible when this is related to the use of technology; and

- simplify procedures as much as possible to minimize use errors.

The QA program must measure each type of activity and task performed by the CE department, including those of the engineers, technologists, clerks, and secretaries.

QA programs for CEDs can be applied to the activities such as: purchase of equipment, incoming inspections, preventive or corrective maintenance, and analysis of incidents and accidents.

The steps to follow when developing a QA program are:

- identify each task and activity that needs to be assessed;

- define the criteria of performance or select the indicators of quality for the activity; and

- compare the criteria to existing standards (if these exist).

4.4.3   EXAMPLES OF QA MEASURES FOR CEDS

Incoming Inspections

Assume the following standard: 100% of new devices must undergo an incoming inspection. If the actual measurement shows that 80% of new devices were inspected by the CED, in the past 6 months, this result is not in compliance with the standard. The next step would be to develop corrective measures. Possible corrective measures could be:

1. to communicate the problem to the purchasing and receiving department to ensure that all deliveries of new equipment, or equipment to be evaluated prior to a purchase, or equipment returned after an external repair be sent to the CED for an incoming inspection;

2. to ensure the CED staff complies and carries out the proper incoming inspections in a timely manner on all devices delivered to the CED; and

3. to repeat this audit in a few months to measure if there is an improvement in the proportion of devices tested when they are delivered to the health care institution.

Preventive Maintenance

Assume the following standard: 100% of critical care equipment will be submitted to a preventive maintenance according to the planned schedule (such as the one recommended by the manufacturer) and 80% of all the other equipment will be tested once per year or after a repair. If the actual measurement is not in compliance, here are some possible corrective measures.

1. Assess the resources available and the workload that the standard as defined above would require.

2. Communicate to health care staff the importance of testing this critical care equipment. Sometimes caregivers are reluctant to let their devices be taken temporarily out of service to be tested; this is especially true in the medical imaging department and the laboratories.

Number of Devices Returning for Repair

Assume the standard to be: no return of equipment within six months of the last repair. Assume the actual measurement was: 1 device was returned 3 times within 6 months. Possible corrective measures are:

1. examine the record and identify past actions;

2. assess what type of intermittent problems could occur and test for these; and

3. ensure another audit is done within 1–3 months to make sure the problem is fixed.

There can be several reasons for this type of problem to occur. One reason can be that the fault is intermittent and so it is difficult to see it when the CED tests it for a repair. Patience is needed to repeat the testing many times to get to a point where the fault occurs.

Remembering one such case in a hospital in Moncton (NB), after such an audit, an electroshock device in psychiatry was sent 3 times within 6 months with the note that it was not discharging properly; the technologist had to repeat the test 25 times before identifying the problem, which he did and repaired. The audit had allowed to search for such incidents in the computerized equipment management system and find this particular problem.

Another reason for such incidences to occur can be due to a 'finger problem'; that is, a user error in using the equipment. User-training would be the solution in this case. Moreover, perhaps simple instructions could be attached to the device for people who do not use it frequently.

Example of a Budget Audit

Assume the standard for the actual expenses of the CED to be within 5% of the approved operating budget. If the actual expenses showed a 10% over expenditure, possible corrective measures are:

1. examine each budget category and expenditure;

2. assess whether this can be corrected or if the hospital needs to adjust the budget; this may be for a case where more repairs and expensive parts were needed than in previous years. Another reason may be that the staff level has not increased along with increased acquisitions of medical devices and so the CED cannot perform all its functions and activities as planned in the QA program. Therefore, the CED manager needs to apply for a staff increase or re-allocate resources to a priorized workload and responsibilities.

Measuring the Satisfaction of Service

The last example is on measuring the satisfaction of service. There are various ways to measure this:

1. send a questionnaire to users in a number of departments, especially those which have a substantial number of devices and who use the CED service more frequently;

2. record the number of complaints (verbal, written);

3. study deficiencies in the accreditation report if any were noted during the accreditation review;

4. assess the average time lag between requests for service and delivery. Sometimes a CED is waiting for a part which is delivered with some delay. A solution in this type of situation is to ensure the owner or user is aware of the reason for the delay. If a repair will take longer than one or two days, then a note could be sent to the department who owns the equipment.

REFERENCES

AAMI.org (n.d.). Available at: http://www.aami.org/publications/standards/60601.html; last accessed May 2013.

ANSI (n.d.). "American National Standards Institute." Available at: www.ansi.org/; last accessed July 2013.

Clemenhagen, CJ (1985). "Quality Assurance in the Hospital- Making it Work." *CMAJ*, 133: editorial.

CSA (n.d.). "Canadian Standards Association." Available at: http://www.csa.ca/cm/ca/en/home; last accessed July 2013.

EMC (n.d.). "Electromagnetic Compatibility." Available at: http://www.medicalelectronicsdesign.com/article/new-emc-requirements-updated-medical-devices-directive; last accessed May 2013.

FDA (n.d.). "Food and Drug Administration." Available at: www.fda.gov/; last accessed July 2013.

Hill, GN, Hill, KT (1981-2005). "Product Liability." Available at: http://legal-dictionary.thefreedictionary.com/Product+Liability; last accessed September 2013.

IEC (n.d.). "International Electrotechnical Commission." Available at: www.iec.ch/; last accessed July 2013.

Institute of Medicine (1985). *Assessing Medical Technologies*. National Academy Press. Washington, D.C.

Keeton, P (1972). "Product Liability and the Meaning of Defect." Available at: http://heinonline.org/HOL/LandingPage?collection=journals&handle=hein.journals/stml-j5&div=12&id=&page=; last accessed September 2013.

"Medical Device – Drug and Health Products." Available at: http://www.hc-sc.gc.ca/dhp-mps/md-im/index-eng.php; last accessed July 2013.

NFPA (n.d.). "National Fire Protection Agency." Available at: www.nfpa.org/; last accessed July 2013.

Roy, OZ (1980). "Summary of Cardiac Fibrillation Thresholds for 60 Hz Currants and Voltages Applied Directly to the Heart." *Med. Biol. Eng. Comput.* 18(5):657-659. DOI: 10.1007/BF02443140.

UL (n.d.). "Underwriters' Laboratory." Available at: http://www.ul.com/global/eng/pages/; last accessed July 2013.

Witters, D (2009). "Medical Devices and EMI: The FDA Perspective." Available at: http://www.fda.gov/MedicalDevices/DeviceRegulationandGuidance/GuidanceDocuments/ucm106367.htm; last accessed July 2013.

OTHER SUGGESTED READING

Dhillon, BS (2008) *Reliability Technology, Human Error, and Quality in Health Care.* CRC Press, Boca Raton, FL.

Dhillon, BS (2000) *Medical Device Reliability and Associated Areas.* CRC Press, Boca Raton, FL. DOI: 10.1201/9781420065596.

Dyro, JF (2004) *Clinical Engineering Handbook.* Ed. Elsevier Academic Press.

Spencer, PC (2002) "Liability Implications for Hospitals of Reprocessing and Reuse of Single-Use Medical Devices." *HR Resources Database*, March. Available at: http://www.longwoods.com/content/16968; last accessed July 2013.

CHAPTER 5

Telemedicine: Applications and Issues

This chapter begins with definitions of telemedicine, telehealth, and telecare, followed by a discussion on drivers of this technology, benefits and limitations, examples of clinical applications, and a brief description of the technology used.

5.1 DEFINITIONS

Telemedicine, in its broadest terms, is medicine delivered at a distance. In more recent terminology, it is meant to represent the transfer of electronic medical data from one location to another. In 1995, telemedicine was defined as the use of telecommunications to provide medical information and services, and in 1999, telemedicine was said to utilize information and telecommunications technology to transfer medical information for diagnosis, therapy, and education. The information can be in the form of images (from pathology, dermatology, or from X-Ray, ultrasound, CT-Scans, MRI); it can be a live video and audio conference between patients and physicians. Video and sound files, patient medical records, and output data from medical devices can all be sent through modern telecommunication networks. The transfer of information or data can be done in real-time or they can be sent at a later time in a block. This can happen in the presence or absence of caregivers, and with or without the presence of the patient (Norris, 2002).

Telehealth is defined as the use of information and communication technologies to transfer health care information for the delivery of clinical, administrative, and educational services. Today, psychologists and community workers are increasingly involved in patient care and some of this can be done from a distance with telehealth.

Telecare utilizes information and communication technologies to transfer medical information for the diagnosis and therapy of patients in their place of domicile. There is a slight difference in the three definitions but overall, the principles remain the same: transmission from a distance, whether it be from a hospital or a patient's home, to a center for diagnosis and medical care (Norris, 2002).

5.2   DRIVERS OF THE TECHNOLOGY

5.2.1   TECHNOLOGICAL DRIVERS

Increased use of computers and of various forms of information technologies in medicine, and the advent of broad network and telecommunication infrastructure in industrialized countries have led to the development and use of telemedicine. Moreover, we live in a technology-led society so there is no surprise to see an increased use of technology for several types of medical services, including the implementation of electronic medical records and telemedicine, among others. The cost, power, and reliability of the technology as well as the implementation of fiber optic networks with broad bandwidth, even in remote and rural areas, have enabled the proliferation of the new technologies and of telemedicine.

5.2.2   NON-TECHNOLOGICAL DRIVERS

Several factors influenced the development of telemedicine services: one major benefit for patients is the access to specialized health services that do not exist in their geographical area; this prevents them from having to drive long distances to access these services. This is particularly important in rural and isolated communities. The weather can be a factor when it is responsible for an absence of transportation that can last days or weeks, which is frequently the case in the Northern Communities of Alaska, the Yukon, Labrador, Newfoundland, and others.

Penal and mental institutions can access health care without moving inmates, an important consideration with the security that this entails in the case of dangerous or violent persons. Telemedicine is also of benefit for travellers on ships or airplanes who experience a medical emergency. For example, British Airways (BA), which flew 30 million people between April 1993 and March 1994 experienced 2078 medical incidents on their flights. They had access 24 hours a day to a physician and could radio the symptoms and other medical data to the medical center in the U.K.. This avoided many unscheduled diversions on long flights, which was of benefit to BA and its passengers (Bagshaw, 1996). In 1999 and 2000, BA added ECG and oxygen monitors, respirators, and portable defibrillators in all their long flight aircrafts. Similarly, at sea, ships can relay patient information through radio links to a medical center which can again minimise diversions to ports not on the schedule (Wootton, 1996).

Another source driving the development of telemedicine was in military applications, to quickly triage casualties and to provide initial emergency treatment in the battlefield. The main types of telemedicine used in military applications were teleradiology and telesurgery.

5.2.3   HOME TELECARE

Home telecare is another important driver of the technology. This arises from the need for patients to download their data, especially in cases where they have a chronic disease such as diabetes. It is also less costly to deliver services to the elderly in their home than to hospitalise them. The elderly frequently have chronic health problems, so regular data transfer of vital signs, glucose, and other data can help manage their health. It is easier for the elderly to be monitored at home or close to where they live, than for them to travel long distances to receive specialized medical services. With the aging population, this will become even more important. The same situation applies to persons with disabilities who would benefit greatly by receiving medical services either in their home or close by.

Another application of telemedicine is in disaster management. Gershnek and Burkle (1999) argue that appropriate telemedicine applications can improve medicine outcomes in cases of disasters; these can be based on previous experience learned from a decade of civilian and military disaster (wide-area) telemedicine deployments. The authors review the history of telemedicine activities in actual disasters and similar scenarios, as well as ongoing telemedicine innovations that may be applicable to disaster situations. They recommend that emergency care providers plan effectively to utilize disaster-specific telemedicine applications to improve future outcomes (Garshnek and Burkle, 1999).

There are few studies about the cost-effectiveness of telemedicine services compared to regular clinical services, but it is intuitive that telemedicine probably costs less than hiring health professionals in areas where the population is sparse. Patients expect to have access to all medical services they need in a timely fashion.

Other drivers would be market development by companies selling equipment, bandwidth, or the telemedicine service itself. Health policy and strategy can also have a major impact on whether the technology gets adopted and used.

5.3   MEDICAL APPLICATIONS OF TELEMEDICINE

Several medical specializations have benefitted from this approach. The most common uses of telemedicine in 1997, in descending order, were in radiology (50%), cardiology (just over 40%), orthopedics (35%), dermatology (32%), and psychiatry (30%) (Congress report, 1997). Globally, in 2009, in developed and developing countries, radiology remained the most common application with 60% of countries offering some teleradiology services, 33% of which were well established services. In the same report, the global figure for telepathology and teledermatology was around 40% with 17% of services well established. For telepsychiatry, the numbers were 24% and 13% (WHO, 2010).

While telemedicine is being used primarily for specialist consultation and second opinions, other applications gained ground such as: the management of chronic illness; emergency and triage; surgical follow-up, diagnostic exam interpretation; and several home health care services. There are also non-clinical uses such as distance learning, grand rounds for physicians, and administrative meetings.

The American Telemedicine Association (ATA) published standards and guidelines that are available at no cost on their website. These apply to several clinical applications such as mental health, teledermatology, diabetic retinopathy, telerehabilitation, telepathology, home care telehealth; new guidelines are also under development (ATA, 2012).

Now we look more closely at examples of telemedicine in a variety of clinical services. In tele-consultation, a patient is examined and interviewed at a distance. This can be in the presence of a local physician, nurse practitioner, or a nurse, and is usually to consult a specialist in a larger center where these do not exist in the region where the patient lives. Aside from providing access to specialists, another result is the enhanced training and education of the physician or nurse in a remote location. Several factors can optimize consultations done by telemedicine. Both sides should agree on the purpose of the consultation, whether it is for making a diagnosis, monitoring the progress of treatment, or to develop the skills of health workers. Both teams should establish the process to be followed and the content of the session and avoid distractions; technology adjustments may be necessary during the consultation, so it is good planning to have a technician accessible nearby to help in these situations, ensuring no time is wasted and that the entire session proceeds smoothly. There should also be some training so that the team using the equipment is familiar with its operation. The delegation of clinical responsibility needs to be formalized and decisions made on the documentation that will be needed.

5.3.1   EXAMPLES OF MEDICAL APPLICATIONS OF TELEMEDICINE

Tele-Radiology

Tele-radiology is one of the most commonly used applications of telemedicine. In 2006, 67% of radiology practices in North America used telemedicine (Steinbrook, 2007). Tele-radiology consists of the transfer of images captured by X-Ray equipment, CT-Scanners, and MRI and Ultrasound equipment. This can be done in real-time (synchronous mode) or in batches (asynchronous mode). Transmissions are usually from a remote location to a larger center, to enable the images to be read and interpreted by radiologists. The approach is also useful for training new radiologists, assisting radiologists in developing countries, diagnosing injured soldiers near or on the battlefield, and performing radiological procedures in space. The images are first digitized, compressed, and sent over a communication network with a fast speed modem; after decompression at the receiving end, the image can be enhanced by magnification, image rotation, or edge enhancement. Care must be taken that the quality of the image does not degrade at each transmission step.

Tele-Surgery

Tele-surgery consists of surgery performed at a distance. It was first developed for military applications, but soon after was applied to civilian situations. The first documented civilian tele-surgery was a gall bladder removal in 2001 between Strasbourg, France, and New York City, a distance of 7000 km (Larkin, 2001). Since then there have been several examples of tele-surgery, such as gall bladder and prostate gland removal, cardiac, thoracic, and urologic surgery. This is usually done with robotic surgery equipment at both ends, with the manipulations from the principal surgeon replicated at the patient location.

Tele-Psychiatry

Tele-psychiatry is where the psychiatrist and the patient meet over teleconference facilities. This approach can be used for the assessment and diagnosis or a variety of mental illnesses, treatment consultations, case conferencing and management, education and supervision, support, forensic and legal assessments, administration and data transfer, research, and psychological testing, among others. Tele-psychiatry has not been shown to be effective in the diagnosis of schizophrenia and is not recommended when dealing with violent or agitated patients. In 2010, in the Province of Ontario, 24 centers were connected to 1 of 3 centers offering tele-psychiatry services for children and youth (MCYS, n.d.). In the U.S., several articles were published in 2013 on tele-psychiatry, which indicates that this practice is still in use and developing (Fortney et al., 2013; Shore, 2013).

Tele-Pathology

Tele-pathology refers to pathology services carried out over a distance. For example, images of tissue captured by a microscope can be transferred to a center where a pathologist can identify if the cells are cancerous or not. This principle is similar to what is used in tele-dermatology where skin rashes or lesions are photographed and sent to a specialist for a diagnosis or for a second opinion. This also includes automated melanoma diagnosis (Wootton and Oakley, 2002). There are also applications in the field of cardiology, obstetrics, pediatric medicine, and many others. There were some interesting telemedicine projects in the U.S. such as the sex assault of children project (available at: http://www.utexas.edu/research/tipi/research/CJA.pdf), Nearly one-fifth of all clinical activity in the U.S. can be attributed to prison telemedicine. Globally, an application that is becoming increasingly important regards the response to disasters.

Tele-Education

Tele-education consists of sessions called grand rounds where patient cases are discussed or presentations are made on new procedures or new treatments. In large cities like New York and London it would make sense to provide continuing education sessions for physicians and interns who work in several medical centers from a single location, as this would save substantial time driving from one place to the other. As mentioned previously, clinical education can also happen during tele-consultation or via the Internet. Public education on preventive medicine and healthy lifestyles can be done through the Internet.

5.4   REPORT TO THE U.S. CONGRESS (1997)

In January 1997, a report on telemedicine was submitted to the U.S. Congress. This included some definitions and the process of evaluation of these types of services. It discussed legal issues such as licensure of physicians offering the services, remuneration issues, safety and standards, telecommunications infrastructure needed to set-up the services, as well as issues concerning privacy, security, and confidentiality. The Federal Evaluation Framework described in the 1997 Report to Congress recommended that evaluations of telemedicine include a description on whether the clinical outcomes were acceptable; the technical acceptability by patients to be assessed at a distance through technology, and by providers using the technology; and the level of integration with the health systems interface. A cost vs. benefits analysis needs to be addressed as well as whether telemedicine can improve access to health care services (Report to Congress, 1997).

5.5   EVALUATION ELEMENTS FOR CLINICAL TELEMEDICINE

When performing an evaluation of a clinical telemedicine service, several elements need to be considered: Do the strategic objectives meet the sponsor's purposes? Regarding the clinical objectives, the evaluation should consider whether the effects intended on the quality of health care, its accessibility and cost meet the expectations. Is the business and project management plan sustainable? There should be a detailed work plan, a schedule, and a realistic budget. The evaluation would normally compare conventional versus telemedicine for specific clinical services to be delivered; the evaluation would assess the technical, clinical, and administrative processes. It should have measurable outcomes and proper documentation.

5.6   THE REPORT ON U.S. TELEMEDICINE ACTIVITY BY THE ASSOCIATION OF TELEMEDICINE SERVICE PROVIDERS (ATSP) IN 1998

The report from the ATSP mentioned that 46 states in the U.S. had some telemedicine services, with California, New York, Texas, and North Carolina being the largest users. The most common use was for consultations and second opinions; second most common was the follow-up of chronic diseases and surgery. There were 41,000 consultations in 1997 and many more were expected in the future. More than 45 specialties using some form of telemedicine were inventoried. A more recent report, *Telemedicine Reimbursement Report of 2003*, can be seen at: http://www.hrsa.gov/ruralhealth/about/telehealth/reimburse.pdf.

An important question remains: Is telemedicine cost-effective? Few studies have yet to answer this question. A literature review of 612 studies revealed in 2002 that only 55 had some form of report on cost-benefit analysis; of the 24 retained for a more comprehensive study, none had included all the criteria judged essential for economic analyses (Whitten et al., 2002). In the year 2000, one review concluded that real-time tele-dermatology was comparable to conventional services; no difference was found in cost effectiveness (Wootton, 2000). But in 2010, Whited published an article on the economic analysis of tele-dermatology and concluded that it was cost effective compared to conventional dermatology services, especially when used in the store-forward mode, an asynchronous model (Whited, 2010).

5.7 REPORT TO THE U.S. CONGRESS (2001)

The report provided an update on telemedicine services in the U.S. and discussed issues such as the lack of reimbursement, legal issues, safety and standards, as well as privacy, security, and confidentiality (Report to Congress, 2001).

5.7.1 BARRIERS AND ISSUES CONCERNING THE IMPLEMENTATION AND USE OF TELEMEDICINE

The list includes but is not limited to the following. Many public and private payees and insurers do not cover telemedicine services; lawyers see the potential for malpractice suits; telemedicine could open the door to out-of-state or out-of-province quacks; people worry about privacy and confidentiality; telemedicine visits must be planned around the physician's and technician's schedule; the possibility of a disclosure of information to partners of patients, in the case of HIV for example.

Paul et al. (1999) reported that end-user and technical training were major barriers, but did not support the quality of the video system reliability, or the perceived inconvenience for physicians to use the equipment as barriers to telemedicine. They argued that a mismatch between the sophistication of the technology and end-user requirements for clinical activities and patient confidentiality and privacy issues were supported as barriers, but how these impacted telemedicine utilization was different than what they had expected. Unsatisfactory sound quality of the telemedicine equipment was identified as a frequent and unexpected barrier to telemedicine utilization rates (Paul et al., 1999).

No doubt there can be substantial benefits to the use of telemedicine: better access to health care with access to specialists not available where they live and a faster diagnosis; improved communication between care-givers; easier continuing education; better access to health information; potentially reduced costs, especially for the patients who would have travel costs, nights away from home, and perhaps need for baby sitting or elderly care help; and reduced time away from work.

There are limitations such as: a lower patient-physician relationship than there is with a face-to-face consultation; the technology is impersonal; or the schedule for fitting in telemedicine may disrupt the organizational workflow. Additional training is needed and protocols need to be developed; there may be an uncertain quality of the health information transmitted and a low rate of utilization which makes the service uneconomical. Another concern is the delay in signal transmission, which is critical in cases of tele-surgery; the guideline requires the delay to be 300 ms or less. This can be a challenge for long distances such as transatlantic transmissions.

Specific issues are: (1) physicians may need a license for each jurisdiction in which they wish to practice; (2) issues of liability—who is responsible for the service, the physician at the distant end or the personnel at the patient end?; (3) many physicians view this as a threat of competition, with services from elsewhere competing with their own; (4) physicians need to plan time for telemedicine consultations or for interpreting the information sent to them between their other duties; and some states are blocking interstate telemedicine in their jurisdiction. There are also questions such as: Do you need a physician at either end? A nurse? A technical expert to operate the equipment?

There are several benefits for physicians: (1) they can attend medical education sessions without having to travel, where this service exists; (2) they can consult a specialist to confirm a patient's condition when this specialization is not available in their geographical area; and (3) they can themselves get referrals if they offer a service that is not available in other places. Some examples in Canada are the following:

- The Hospital for Sick Children in Toronto opened a telemedicine clinic with Thunder Bay and with four other health regions in Northern Ontario.

- The University of Ottawa Heart Institute offers telemedicine cardiac care to a hospital in the small town of Pembrooke (Ontario).

- Parry Sound and six smaller centers receive tele-psychiatry services from Toronto.

- The oldest telemedicine service in Canada (1977) began with Memorial University and its Health Sciences Centre in St-John's, offering telemedicine services to several remote parts of Newfoundland.

- St-John Regional Hospital in New Brunswick interprets X-Ray images sent from Grand Manan Island (NB), thus reducing expensive and long trips from the island to the mainland.

5.8   ETHICAL AND LEGAL ASPECTS

All countries and states using telemedicine need to develop policies that affect access, delivery, reimbursement, licensure of users, maintenance of confidentiality and privacy. The policies must clarify what rights of access patients in a certain jurisdiction will have. This of course depends on the level and quality of services offered. There must be policies to ensure the protection of data gathered by the means of telemedicine. The expected duty of care and standard of care should be the same as what is expected from a conventional clinical visit. Physicians are as vulnerable to malpractice suits, whether they see patients in their office or via telemedicine. The suitability of the equipment in transmitting an acceptable level of quality of images and data is important. A process must be in place in case of failure of the equipment; perhaps back-up equipment would be made available or the patient re-scheduled, or if more urgent, then the patient may be referred to another center that can offer the service.

Physicians must be licensed for the various jurisdictions where they will offer their services by telemedicine. The facility must be accredited and a schedule established for the re-imbursement of the services. Intellectual property rights must be clearly identified as telemedicine involves more than one center and team. Finally, it is critical to ensure the confidentiality, privacy, and security of patients and their data. For example, it may not be clear to the patient, who is in a room at one end, who is present at the other end, whether seen by the camera or not; everyone in the room must be identified.

5.9   TECHNICAL REQUIREMENTS

T1 lines that transmit at a speed of 1.54 Mbps would support imaging modalities. Multiplexing can be used to fractionate a T1 line for use in television and to transmit medical images at the same time. Interactive video is a useful modality for consultations. Leased ISDN telephone lines (Integrated Services Digital Network) have twice the capacity of regular lines. High speed fiber optic lines are excellent for maintaining the quality and speed of transmission.

REFERENCES

ATA (2012). "The American Telemedicine Association." Available at: http://www.american-telemed.org/practice; last visited July 2013.

Bagshaw, M (1996). "Telemedicine in British Airways." *J. Telemed Telecare*, 2 (suppl): 36-38.

Choi, YB, Krause, JS, Seo, H, Capitan, KE, Chung, K (2006). "Telemedicine in the USA: Standardization through Information Management and Technical Applications." *IEEE Comm. Mag.* April: vol 44(4): 41-48. DOI:10.1109/MCOM.2006.1632648.

Fortney, JC, Pyne, JM, Mouden, SB, Mittal, D, Hudson, TJ, Schroeder, GW, et al. (2013). "Practice-Based Versus Telemedicine-Based Collaborative Care for Depression in Rural Federally Qualified Health Centers: A Pragmatic Randomized Comparative Effectiveness Trial." *Amer. J. Psychiat.* April, 4(170): 414-425. DOI:10.1176/appi.ajp.2012.12050696.

Garshnek, V, Burkle, FM (1999). "Telemedicine and Telecommunications to Disaster Medicine Historical and Future Perspectives." *J. Am. Med. Inform. Assoc. (JAMIA)* Jan-Feb, 6(1): 26–37.

Larkin, M (2001). "Transatlantic, Robot-Assisted Telesurgery Deemed a Success." *Lancet*, 358 (9287): 1074.

MCYS (n.d.). "Ministry of Children & Youth Services." Available at: http://www.children.gov. on.ca/htdocs/English/news/factsheets/05102007.aspx; last accessed July 2013. DOI: 10.1016/S0140-6736(01)06240-7.

Paul, DL, Pearlson, KL, McDaniel, Jr. RR (1999). "Assessing Technological Barriers to Telemedicine:Technology-Management Implications." *IEEE Trans. Eng. Manag.*, Aug., 46 (3): 279-288.

Report to Congress (1997). Available at: http://www.ntia.doc.gov/legacy/reports/telemed/index. htm; last accessed July 2013. [This report also contains the 1996 report from the Institute of Medicine on Telemedicine.]

Report to U.S. Congress (2001). Available by request at: http://ask.hrsa.gov/detail_materials. cfm?ProdID=402; last accessed July 2013.

Shore, JH (2013). "Telepsychiatry: Videoconferencing in the Delivery of Psychiatric Care." *Amer. J. Psychiat.*, March, 3(170): 256-262.

Steinbrook, R (2007). "The Age of Teleradiology." *New Engl. J. Med.*, 357:5-7. DOI: 10.1056/ NEJMp078059.

Whitten, PS, Mair, FS, Haycox, A, May, CR, Williams, TL, Hellmich, S (2002). "Systematic Review of Cost Effectiveness Studies of Telemedicine Interventions." *BMJ*. June 15, 324(7351): 1434–1437.

Whited, JD (2010). "Economic Analysis of Telemedicine and the Teledermatology Paradigm." *Telemed e-Health*, March, 16(2): 223-228.

Wooton, R (1996). "Telemedicine—A Cautious Welcome." *BMJ* 313:1375-1377.

Wootton, R, Bloomer, SE, Corbett, R, Eedy, DJ, Hicks, N, Lotery, HE, et al. (2000). "Multicentre Randomised Control Trial Comparing Real Time Teledermatology with Conventional Outpatient Dermatological Care: Societal Cost-Benefit Analysis." *BMJ*. May 6; 320(7244): 1252–1256.

Wootton, R, Oakley, A (2002). *Teledermatology*. RMS Press.

World Health Organization, (2010). "Telemedicine: Opportunities and Developments in Member States. Report on the Second Global Survey on e-health." Available at: http://www.who.int/goe/publications/goe_telemedicine_2010.pdf; last accessed July 2013.

OTHER SUGGESTED READING

Norris, AC, (2002). *Telemedicine and Telecare.* Wiley. DOI: 10.1002/0470846348.

Toader, E, Damir, D, Toader, IA, (2011). "Ethical and Legal Issues Related to the Clinical Application of Telemedicine." *Proc. Third Intern. Conf. E-Health and Bioengineering.* Romania.

Xiao, Y, Chen, H, (2008). *Mobile Telemedicine – A Computing and Networking Perspective.* CRC Press. Taylor and Francis Group.

CHAPTER 6

Impact of Technology on Health Care and the Technology Assessment Process

6.1 IMPACT OF TECHNOLOGY ON HEALTH CARE

Chaudry et al. (2006) did a systematic review of the impact of health information technologies (IT) on the quality, efficiency, and cost of medical care; 25% of the information came from four institutions which had implemented internally developed systems, most of which were related to decision support and electronic health records. The authors concluded that, for these four institutions, the quality and efficiency of medical care were increased, but it was unclear whether cost was impacted and whether the results could apply to other institutions. Of 257 studies the authors reviewed, the primary domain of improvement was in preventive health. These authors found three benefits that were considered as a positive impact: an increased adherence to guideline-based care; enhanced surveillance and monitoring; and a decrease in medication errors. In terms of increased efficiency, they found that there was a decrease in the utilization of care, but the time used in providing care showed mixed results. The availability of empirical cost data was limited (Chaudry et al., 2006).

There was an increased use of technology for diagnosis, therapy, and monitoring during the last two decades of the 20th century and the complexity of the devices increased over this time period. The development of the Internet had an impact on the amount of information available to health care personnel, but has led to information overload. However, there are both positive and negative aspects to a greater connectivity. On the negative side, too much information can create confusion, distraction, and even errors. On the positive side, the Internet provides access to journals and medical information quickly, enabling caregivers to keep up to date. Consultations for a second opinion or for an expert opinion are faster and more efficient. Some Decision-Support Systems (DSS) can digest some of the information and provide knowledge that can help health care workers make better and faster decisions in fast-paced environments. (See Part II for a discussion on the impact of technology on the reduction of medical errors.) It is important to take into consideration that an increased use of technology to sustain life also raises the issue of the quality of life, an ethical question.

There is an economic aspect to this discussion. Expenditures on health care have been increasing at high rates in the past two decades. For example, in the U.S., health care expenditures were 13.5% of GDP in 1997, at a time when other western nations were at 10% (Germany, France, and Switzerland). Total U.S. health expenditures rose from $888 billion in 1993 to $1,425 billion in 2001; it grew to 2.6 trillion in 2012. The health expenditure per capita was $8232.90 in 2012, which was twice the OECD average. In Canada, the health expenditure per capita was $4607.80 in the same year. It is not simple to determine the impact of technology on health care costs. Technology has been expensive to acquire, but it also has had some impact on the reduction of costs. It is perhaps more important to focus on ways that the acquisition of technology has changed the way health care is delivered and received.

Some benefits of medical technologies are: Most technologies today have some form of automation which can save time for caregivers who can then spend more of their time on direct care to patients. Another benefit is that patients have more ways to keep aware of their chronic health conditions with home monitoring, or downloading data to heath care centers; thus, they may adhere to lifestyles and medication regimens that could delay the progression of the disease. It is not difficult to imagine that, without this information, patients would have had multiple hospitalizations in past eras each time their condition worsened. The reduced use of hospital beds by these patients likely represents a major cost reduction for health care. Technology has, in many cases, enabled caregivers to make an earlier diagnosis and thus begin therapy at early stages of a disease, which would have some impact on outcomes and on the quality of health care. Colás et al. (2010) made this point for one of the most expensive chronic condition: cardiac insufficiency or heart failure. These authors argue that using implantable devices to stabilize their condition, and performing the regular follow-up by telemedicine, provides major savings to the health care system in terms of physician time, hospital capacity, and transportation costs for the patient. On the impact of health information technologies on chronic care management, Marchibroda (2008) wrote:

> The introduction of health IT, including electronic health records and health information exchange, holds great promise for addressing many of the barriers to effective chronic care management, by providing important clinical information about the patient when it is needed, and where it is needed, in a timely, secure fashion. Having information from the care delivery process readily available through health IT and health information exchange at the national, state, and local levels supports key components of the chronic care management process, including those related to measurement, clinical decision support, collaboration and coordination, and consumer activation (**Marchibroda, 2008**).

A U.S. report by the National Center for Health Statistics (NCHS) shows, among many other variables, the actual amount invested in durable medical equipment between 1960 and 2009, and the proportion that this cost represents of the total health expenditures in the US. In 1960, 0.7 billion dollars were spent on medical equipment (2.6% of the total health expenditures). Although the actual amount in dollars increased each year, reaching 34.9 billion in 2009, the proportion of the equipment acquired over the total expenditures decreased over the years to reach a low value of 1.4% in 2009, the last year for which this data is available (NCHS, 2011).

However, the complexity of medical equipment continues to grow, and so is the disproportionate relationship between initial cost and on-going support costs; the latter is expected to continue to increase in the years to come. Design innovation, especially with regards to autonomic systems, will require careful management. Clinical engineers can play a major role, not only in the management of the technology, but they may also be able to influence some of the designs.

6.2   THE HEALTH TECHNOLOGY ASSESSMENT PROCESS

Since the middle of the twentieth century, medical equipment has become an essential part of the provision of health care services in industrialized countries and even in many developing countries. With the proliferation of medical devices flooding the market, and the spiraling increase of health care expenditures, hospitals and other health care facilities have to make informed decisions on the acquisition of new technologies. To help them in this process, several nations have created health technology assessment (HTA) centers.

In addition to these HTA centers, there are groups of clinical and biomedical engineers and not-for-profit organizations performing the evaluation of medical devices; these groups compare different brands and various designs which perform a similar medical function. Important questions are: What devices should be tested, and when? What tests are needed? Who should do the testing? An effective approach is to test the devices in a human factors engineering laboratory which can simulate the real environment and not only test the devices themselves but also the interaction of users with the equipment (Easty et al., 2009). For example, Chan et al. (2011) evaluated the task efficiency, usability, and safety of three order set formats: their hospital's planned CPOE (Computerized Physician Order Entry) order sets, computer order sets based on user-centered design principles, and existing pre-printed paper order sets. The participants were 27 physicians, residents, and medical students. The measures were completion time (efficiency), requests for assistance (usability), and errors in the submitted orders (safety). The User-Centered Design format was more efficient and usable than the CPOE test format, even though training was provided for the latter. The authors concluded that application of user-centered design principles can enhance task efficiency and usability, thus increase the likelihood of successful implementation (Chan et al., 2011). This example shows how this type of testing can be effective in selecting the best technologies for specific functions and their potential users.

In addition to the existence of national and regional organizations in many countries that provide medical technology assessments, there are also international agencies supporting the advancement of HTA on the global stage such as Health Technology Assessment international (HTAi) and the International Network of Agencies in Health Technology Assessment (INAHTA).

6.2.1   WHAT IS A HEALTH TECHNOLOGY ASSESSMENT (HTA)?

Technology assessment originated in the 1960s and was meant principally to instruct policy-making by governments on health technology investments. Goodman provides a definition of TA from an Institute of Medicine Publication in 1985:

> We shall use the term assessment of a medical technology to denote any process of examining and reporting properties of a medical technology used in health care, such as safety, efficacy, feasibility, and indications for use, cost, and cost-effectiveness, as well as social, economic, and ethical consequences, whether intended or unintended (**Goodman 2004, in Institute of Medicine, 1985**).

Other definitions are listed in Goodman (2004): "Technology assessment is a form of policy research that examines short- and long-term social consequences (societal, economic, ethical, and legal) of the application of technology." For his part, Banta (1993) wrote: "The goal of technology assessment is to provide policy-makers with information on policy alternatives" (Banta, 1993).

The U.K. National Health Service R&D Health Technology Assessment Programme (2003) defines HTA:

> Health technology assessment considers the effectiveness, appropriateness and cost of technologies. It does this by asking four fundamental questions: Does the technology work, for whom, at what cost, and how does it compare with alternatives?

The U.S. Congress Office of Technology Assessment (1994) provided the following definition: "Health technology assessment ... is a structured analysis of a health technology, a set of related technologies, or a technology-related issue that is performed for the purpose of providing input to a policy decision."

In Canada, the Canadian Agency for Drugs and Technology Assessment (CADTH) writes: "Decisions about which medical devices and drugs to use are crucial to the quality and sustainability of health care in Canada. Access to evidence-based information is key to making informed decisions that harness the benefits of technology while getting the best value from every health dollar. CADTH provides decision-makers with the evidence, analysis, advice, and recommendations they require to make informed decisions in health care."

The International Network of Agencies for Health Technology Assessment (INAHTA) writes: "Health technology assessment (HTA) is the systematic evaluation of properties, effects, and/or impacts of health care technology. It may address the direct, intended consequences of technologies as well as their indirect, unintended consequences. Its main purpose is to inform technology-related policymaking in health care. HTA is conducted by interdisciplinary groups using explicit analytical frameworks drawing from a variety of methods. In addition to HTA being used for informing governments on health policy, it is also very useful for health care facilities planning to invest substantial funds into new technologies. It is prudent for these health care institutions to find out as much as they can about the new technology they are thinking of acquiring.

A technology assessment consists of several aspects: technical, economic, medical effectiveness, and safety. The usual approach is to compare the various ways of accomplishing a clinical function. Take, for example, the measurement of blood glucose in the home environment. The following alternatives exist:

1. small glucose monitors where patients prick their finger, place the blood on a strip, which is then read by a monitor;

2. there is a watch that patients can wear to get a continuous reading of their glucose level. Advantages and disadvantages of each approach, their accuracy, the total capital cost and the cost per test should be compiled; and

3. finally, the life cycle of the device and the safety for patients and staff need to be considered.

HTA centers usually separate the work they do into two types of assessments: drugs and technologies. In Canada, the national HTA is called "Canadian Agency for Drugs and Technologies in Health"; in the U.S., the Office for Technology Assessment (OTA), which published over 750 studies, was closed in 1995. However, there is a strong voice for re-funding this type of organization; Sadowski (2012) wrote: "there has been vocal support by many prominent scholars and politicians to either re-fund it or establish a similar method of technology assessment."

There is a difference between evaluation and assessment. ECRI has over many years provided evaluations of a variety of medical devices. For example, they can provide a comparison of a number of electrosurgical generators. This would be similar to the Consumer Reports comparing trucks, cars, etc. An assessment as carried out by HTA centers would look at different manners of performing a clinical function rather than comparing similar equipment from different manufacturers as described in the example below for kidney stone removal.

In this example of a medical technology assessment, it is appropriate to compare various ways to remove kidney stones:

1. a traditional surgery;

2. a catheter introduced though the urethra to the kidney and ultrasound signal to blast the stones; and

3. a newer approach called lithotripsy; in this case, there are at least three different types of technologies to accomplish the task. Conventional surgery requires several days of hospitalization and the risk for an infection is high.

The ultrasound approach is common, but sometimes the stones are located in pockets not reachable or visible by this method. Lithotripsy does not require hospitalization and is non-invasive, as the wave generated to blow up the stone is applied though the skin. This equipment is quite expensive and is likely justified for a health care facility serving a population of a million people or more. Detailed costs per procedure need to be identified. Serviceability is another aspect to consider.

Types of organizations that carry-out HTAs are numerous: regulatory agencies, government and private sector payers, managed care organizations, health professions organizations, standards setting organizations, hospitals and health care networks, group purchasing organizations, patient and consumer organizations, government policy research agencies, private sector assessment/policy research organizations, academic health centers, biomedical research agencies, health product companies, venture capital groups, and other investors (Goodman 2004).

A framework for HTA was offered by the European Collaboration for Health Technology Assessment (Busse, 2002, in Goodman, 2004) where the following steps are usually followed: the process begins with a submission of an assessment request or an identification of an assessment need; a prioritization is done as to which HTAs are more urgent; the work is commissioned by selecting the proper group to carry-out the HTA.

An assessment follows these steps:

- Definition of the policy question(s);

- Elaboration of the HTA protocol;

- Collecting background information and determination of the status of the technology

- Definition of the research questions;

- Sources of data, appraisal of evidence, and synthesis of evidence for each of safety, efficacy/effectiveness, psychological, social, ethical, organizational, professional, economic considerations.

A draft is written on the discussions, conclusions, and recommendations, followed by an external review, and publishing of the final HTA report with an executive summary of the report. The next step is dissemination of the report. Later on, there may be an update needed of this particular HTA. Several countries follow a similar process for their HTAs.

In conclusion, there are several centers performing HTAs around the world, so a literature review prior to planning a new HTA is essential. Perhaps HTAs that have been done are now obsolete and in need of a review or update, or an HTA may be more appropriate for a particular environment than another. In general, HTAs are useful and can save much effort and/or a waste of investment for future acquisitions of medical technologies.

REFERENCES

Banta, HD, Luce, BR (1993). *Health Care Technology and Its Assessment: An International Perspective.* New York, NY: Oxford University Press.

Busse, R, Orvain, J, Velasco, M, et al. (2002). "Best practice in undertaking and reporting health technology assessments." *Int. J. Technol. Assess. Health Care*, 18: 361-422. DOI: 10.1017/S0266462302000284.

Canadian Agency for Drugs and Technology Assessment (CADTH). Funded by Canada's federal, provincial, and territorial governments, CADTH is an independent, not-for-profit agency that delivers timely, evidence-based information to health care leaders about the effectiveness and efficiency of health technologies. http://www.cadth.ca/en/cadth.

Chan, J, Shojania, KG, Easty, AC, Etchells, EE (2011). "Does User-Centred Design Affect the Efficiency, Usability, and Safety of CPOE Order Sets?" *J.A.M.I.A.* May, 18(3): 276-281. DOI: 10.1136/amiajnl2010000026.

Chaudhry, B, Wang, J, Wu, S, Maglione, M, Mojica, W, Roth, E, et al. (2006). "Systematic Review: Impact of Health Information Technology on Quality, Efficiency, and Costs of Medical Care." *Ann. Intern. Med.* 144: E-12-E-22. DOI: 10.7326/0003-4819-144-10-200605160-00125.

Colás, J, Guillén, A, Moreno, R (2010). "Innovation in Health Care technology: Is it part of the problem or part of the solution? eHealth gives the answer." *Proc. IEEE/EMBS Conf.* Aug.-Sept. : 1057-1060. DOI: 10.1109/IEMBS.2010.5627640.

Easty, A. C., Cafazzo, J. A., Chagpar, A (2009). "Improving Safety in Healthcare through the Establishment of a Healthcare Human Factors Team." In *World Congress on Medical Physics and Biomedical Engineering*, September 7-12, Munich, Germany, Springer Berlin Heidelberg: 324-327. DOI: 10.1007/9783642038853_90.

Goodman, CS (2004). "HTA 101: Introduction to Health Technology Assessment." Available at: http://www.nlm.nih.gov/nichsr/hta101/ta101_c1.html; last visited in May 2013.

Institute of Medicine (1985). *Assessing Medical Technologies.* National Academy Press. Washington, D.C.

International Network of Agencies for Health Technology Assessment (INAHTA) . http://www.inahta.org/GO-DIRECT-TO/Members/

Marchibroda, JM (2008). "The Impact of Health Information Technology on Collaborative Chronic Care Management." *J. Managed Care Pharm. Suppl.*, March, 14(2): S3-S11.

NCHS (2011). "Table 128." Available at: http://www.ncbi.nlm.nih.gov/books/NBK98779/table/trendtables.t128; last accessed July 2013.

Sadowski, J (2012). "Non-Partisan Advice Needed by Congress." *The Atlantic*. October 26. Available at: www.ota.fas.org; last accessed September 2013.

"U.K. National Health Service R&D Health Technology Assessment Programme" (2003). Available at: http://www.ncchta.org/about/index.shtml; last accessed June 1, 2003.

U.S. Congress, Office of Technology Assessment (1994). "Protecting Privacy in Computerized Medical Information." U.S. Government Printing Office. Washington, D.C.

World Health Organization,(n.d.). "Technology Assessment." Available at: http://www.who.int/medical_devices/assessment/en/; last accessed September 2013.

Author Biography

Monique Frize is a Distinguished Professor at Carleton and Professor Emerita at University of Ottawa. For 18 years (1971-1989) she was a hospital biomedical engineer and has been a professor in electrical and biomedical engineering since 1989. Monique Frize has published over 200 journal and conference proceedings papers on artificial intelligence in medicine, infrared imaging, ethics, and women in engineering and science. She is a Fellow of IEEE (2012), the Canadian Academy of Engineering (1992), Engineers Canada (2010), Officer of the Order of Canada (1993), and recipient of the 2010 Gold Medal from Professional Engineers Ontario and the Ontario Society of Professional Engineers. She has received five honorary doctorates in Canadian universities since 1992. Her book, *The Bold and the Brave: A History of Women in Science and Engineering* was released by the University of Ottawa Press in November 2009; *Ethics for Bioengineers* was published by Morgan & Claypool (2011); and her new book, *Laura Bassi and Science in 18th Century Europe: The Extraordinary Life and Role of Italy's Pioneering Female Professor*, was released by Springer in July 2013.

Printed in the United States
by Baker & Taylor Publisher Services